Lecture Notes in Mechanical Engineering

Lecture Notes in Mechanical Engineering (LNME) publishes the latest developments in Mechanical Engineering—quickly, informally and with high quality. Original research reported in proceedings and post-proceedings represents the core of LNME. Volumes published in LNME embrace all aspects, subfields and new challenges of mechanical engineering. Topics in the series include:

- Engineering Design
- Machinery and Machine Elements
- Mechanical Structures and Stress Analysis
- Automotive Engineering
- Engine Technology
- Aerospace Technology and Astronautics
- Nanotechnology and Microengineering
- Control, Robotics, Mechatronics
- MEMS
- Theoretical and Applied Mechanics
- Dynamical Systems, Control
- Fluid Mechanics
- Engineering Thermodynamics, Heat and Mass Transfer
- Manufacturing
- Precision Engineering, Instrumentation, Measurement
- Materials Engineering
- Tribology and Surface Technology

To submit a proposal or request further information, please contact the Springer Editor of your location:

China: Dr. Mengchu Huang at mengchu.huang@springer.com
India: Priya Vyas at priya.vyas@springer.com
Rest of Asia, Australia, New Zealand: Swati Meherishi at swati.meherishi@springer.com
All other countries: Dr. Leontina Di Cecco at Leontina.dicecco@springer.com

To submit a proposal for a monograph, please check our Springer Tracts in Mechanical Engineering at http://www.springer.com/series/11693 or contact Leontina.dicecco@springer.com

Indexed by SCOPUS. The books of the series are submitted for indexing to Web of Science.

More information about this series at http://www.springer.com/series/11236

Elisabetta Ceretti · Tullio Tolio
Editors

Selected Topics in Manufacturing

AITeM Young Researcher Award 2019

Editors
Elisabetta Ceretti
Dipartimento di Ingegneria Meccanica e Industriale
Università di Brescia
Brescia, Italy

Tullio Tolio
Mechanical Engineering Department
Politecnico di Milano
Milano, Italy

ISSN 2195-4356 ISSN 2195-4364 (electronic)
Lecture Notes in Mechanical Engineering
ISBN 978-3-030-57731-5 ISBN 978-3-030-57729-2 (eBook)
https://doi.org/10.1007/978-3-030-57729-2

This Springer imprint is published by the registered company Springer Nature Switzerland AG
The registered company address is: Gewerbestrasse 11, 6330 Cham, Switzerland

Scientific Committee

Prof. Elisabetta Ceretti (President), University of Brescia
Prof. Umberto La Commare (Vice-President), University of Palermo
Prof. Domenico Umbrello (Secretary), University of Calabria
Prof. Arianna Alfieri, Politecnico di Torino
Prof. Antonello Astarita, University of Naples Federico II
Dr. Giacomo Bianchi, STIIMA-CNR (National Research Council of Italy)
Prof. Giuseppe Casalino, Politecnico di Bari
Prof. Antoniomaria Di Ilio, University of L'Aquila
Prof. Barbara Previtali, Politecnico di Milano
Dr. Antonio Scippa, University of Firenze

Among the contributions in this book, the *AITeM Young Researcher Award 2019—Edoardo Capello Prize* has been assigned to Dr. Giovanna Rotella for the work *Surface modifications induced by roller burnishing of Ti6Al4V under different cooling/lubrication conditions* by G. Rotella and L. Filice.

Preface

Manufacturing can be easily considered the backbone of the European economy with a total of 2.1 million enterprises employing near 30 millions of people (14.2% of the total European workforce) and generating EUR 1710 billions of value added in 2014 (26 % of the European value added).

The competitivity and resilience of the European manufacturing sector strongly ground on the technical leadership and ability to handle complexity. Manufacturing is today a complex and highly interconnected value creation process ecosystem pursuing high-value-added products to compete globally.

Manufacturing is evolving continuously, taking advantage of emergent technologies and establishing interrelations with many scientific and technological areas, e.g., computer science, materials research, microelectronics, and biosciences, moving its scope beyond simple product fabrication.

Italy is one of the main actors in the European manufacturing sector, being the second largest manufacturing economy in Europe. Moreover, Italy has also a prominent role as a technology provider for manufacturing, with the machinery and equipment sector having the largest share of the Italian exported goods in the last years. Research and innovation also constitute a very relevant area for both Italian universities and companies, covering different and heterogeneous sectors, ranging from manufacturing technologies, processes, equipment, systems, as well as strategical and economical aspects. Italian partners, in fact, reached the second position in terms of research efforts in European H2020 projects in manufacturing.

AITeM, the Italian Association of Manufacturing, funded in 1992, is an organization involving academics, researchers, and industrialists whose main interest is in manufacturing. Since the foundation, AITeM is the cultural and technological reference for manufacturing and production systems in Italy, whose aim is promoting research in manufacturing through events, collaborations, research and industrial projects, as well as disseminating the research and innovation culture to the general public. AITeM counts today about 300 associates coming from all the universities in Italy, as well as 30 industrial groups. AITeM is also the largest network of research laboratories in Italy in the area of manufacturing.

This volume contains a set of selected contributions proposed by young AITeM associates and refers to a wide range of scientific and technological areas: additive manufacturing, advanced and unconventional machining and processes, material removal processes, foundry and forming, tools and machine tools, assembly/disassembly, joining materials and material properties, quality metrology and material testing, manufacturing systems engineering, sustainable manufacturing, smart manufacturing and cyber-physical systems, education in manufacturing and human factors, and industrial applications.

This book addresses the multifaceted nature of the research in manufacturing, capturing, and interconnecting different scientific and technological areas. It also provides a picture of the vitality of the Italian research community looking toward the future.

Milan, Italy — Tullio Tolio, President of AITEM

Brescia, Italy — Elisabetta Ceretti, President of the Scientific Committee

May 2020

Contents

Micro-milling of Selective Laser Melted Stainless Steel

Andrea Abeni, Paola Serena Ginestra, and Aldo Attanasio

Abstract This paper deals with micro mechanical machining process of 17-4 PH stainless steel samples fabricated by selective laser melting. An analysis of the material removal behaviour during micro-milling operations for the selection of the optimal feed rate value was performed on 17-4 PH additive manufactured samples studying the variation of the specific cutting force as a function of the feed per tooth. The transition from shearing to ploughing regime was analysed by considering the variation of the specific cutting forces. The minimum uncut chip thickness was calculated to identify the transition between the cutting regimes (shearing, ploughing or their combination) that affects the final product quality in terms of surface integrity and dimensional accuracy. Moreover, the surface roughness and the burr extension were analysed as a function of the feed rate.

Keywords Selective laser melting · Micro machining · Minimum uncut chip thickness

1 Introduction

Differently from conventional machining processes, Additive Manufacturing (AM) processes produce parts with complex shape by material addition. Depending on the material charging method, the AM techniques of metals can be classified in: powder bed fusion, direct energy deposition and wire fed systems. Selective Laser Melting

A. Abeni (✉) · P. S. Ginestra · A. Attanasio
Department of Mechanical and Industrial Engineering, University of Brescia, 25123 Brescia, BS, Italy
e-mail: andrea.abeni@unibs.it

P. S. Ginestra
e-mail: paola.ginestra@unibs.it

A. Attanasio
e-mail: aldo.attanasio@unibs.it

E. Ceretti and T. Tolio (eds.), *Selected Topics in Manufacturing*, Lecture Notes in Mechanical Engineering,
https://doi.org/10.1007/978-3-030-57729-2_1

(SLM) is the most promising powder bed fusion process where a product is obtained by the selective melting of metal powders by a laser source. SLM allows the fabrication of products characterized by high structural integrity. On the other hand, the surface finish is inadequate and with high variability that in some cases can affect the technical properties and compromise the required tolerances [1]. The final properties of SLM metals are still under study due to the presence of uncontrolled porosities, defects and poor surface finishing states [2]. The poor surface quality of the SLM components resulting from a high surface roughness is mostly due to the partially melted powder on the outer surface of the manufactured parts collected during the building process [3]. Post processing is therefore needed for an improvement of the surface finishing and mechanical properties of the final parts.

Among the traditional processes used to achieve high precision on 3D components, micro milling is one of the most convenient micro manufacturing processes in terms of volume and cost ratio [4, 5]. Micro milling can be utilized to mechanically remove materials using micro tools to obtain complex micro-size features on a wide variety of engineering materials. However, the efficiency of micro milling introduces critical issues due to the miniaturization of parts and tools that requires a deep understanding and optimization of the process. Micro machining operations are characterized by a chip thickness comparable in size to the cutting edge radius of the mill. The increases in cutting energy and forces as the undeformed chip thickness decrease is one of the most significant size effects of micro milling. In particular, when the uncut chip thickness is lower than a minimum value (i.e. minimum uncut chip thickness), the cutting process is characterized by an elasto-plastic deformation of the material known as ploughing. This cutting regime is does not correspond to a correct chip formation. Thus, the Minimum Uncut Chip Thickness (MUCT) has been identified as the undeformed chip thickness at which the transition from ploughing to shearing occurs causing a significant variation of the normalized cutting energy and forces [6]. In order to increase productivity and improve the machined part quality during micro milling, the ploughing mechanism has to be understood. Moreover, the ploughing process has a direct impact on the final surface roughness of the treated material influencing the achievable accuracy of the finished AM components. Furthermore, the analysis of the dominant deformation regime during micro milling operations involves the cutting force measurement. The effects of the tool run-out must be considered to quantify the loads imbalance on the tool flutes during the process [7].

The objective of this paper is to study the material removal behaviour of 17-4 PH stainless steel parts produced by SLM and post processed by micro milling. The proposed analysis is based on the realization of microchannels with 800 μm width by using coated tungsten carbide micro end mills on SLM steel samples. The cutting force was acquired at a high sampling rate in order to avoid any aliasing effects. In particular, the cutting force has been analysed as a function of the feed per tooth in order to identify the occurring transition from ploughing to shearing. A proper analytical model to take into account the tool run-out effects while calculating the specific cutting forces was applied.

Moreover, the roughness (Ra) and the burrs dimension of the microchannel were analysed to relate the surface finishing to the material deformation mechanism during the process.

2 Experimental Procedure

In this section, the production and machining operations of the SLM 17-4 PH stainless steel samples are reported. The manufacturing parameters and the experimental plan followed for the micro milling tests are defined and the model used for the evaluation of the specific cutting force in presence of the tool run out is described.

2.1 *Sample Production*

The SLM samples were produced using the laser based powder bed fusion machine ProX 100 (3D System). The geometry of the samples was designed to allow the positioning of the samples on the load cell Kistler© 9317C. The samples were printed as squares with a side equal to 25 mm and a thickness equal to 5 mm with four holes with a diameter of 4.20 mm on the corners of the squares.

The chemical composition of the 17-4 PH stainless steel powder used as printing material is reported in Table 1.

The absence of impurities is necessary to avoid negative effects of embrittlement. Therefore, the laser process is carried out in a Nitrogen atmosphere with a controlled O_2 content less than 0.1 vol.%. The optimized parameters for the SLM of 17-4 PH steel of the process are reported in Table 2.

The as built samples were subjected directly to the micromachining tests without heat treatment.

Table 1 Chemical composition of 17-4 PH stainless steel powder

17-4 PH	C	Cr	Ni	Cu	Mn	Mo	Nb	Si
Wt (%)	<0.07	16.71	4.09	4.18	0.8	0.19	0.23	0.53

Table 2 Process parameters used in the SLM process

Process parameter	Value
Laser power (W)	50
Spot diameter (μm)	80
Scan speed (mm/s)	300
Hatch spacing (μm)	50
Layer thickness (μm)	30

2.2 Micro-milling Tests

The machining tests were carried out fabricating twenty channels by using a constant cutting speed and twenty different feed per tooth (f_z) values. Once the cutting forces were acquired and normalized, the ploughing-shearing transition was determined through the MUCT quantification. Moreover, the roughness and the burrs were measured and the data were evaluated as a function of the parameter f_z.

The cutting tests were performed on a five axis Nano Precision Machining Centre KERN Pyramid Nano equipped with a Heidenhain iTCN 530 numeric control. The loads generated by the interaction between tool and workpiece were measured through a force acquisition system, as reported in [8]. The precision of the load cell, the bandwidth and the sampling rate of the measurement system are adequate for capturing forces in micro milling [9]. The samples were constrained to the load cell through four bolts. The load cell was blocked to the machine work table. The experimental procedure consisted in two different milling operations: (i) a roughing to prepare a planar surface on the workpiece and (ii) micro slot machining, executed varying the fz at each test. The force acquisition was performed during the micro slot machining. The roughing was performed through four identical consecutive passes with a depth of cut of 100 μm for each step. A four-flutes flat-bottom mill with a nominal diameter of 6 mm was employed to prepare the samples setting a cutting speed equal to 40 m/min and a feed per tooth of 10 μm/tooth. The microchannels were produced using a coated two flutes micro mill with a nominal diameter of 0.8 mm. The actual tool geometry was acquired using a confocal microscope (Hirox RH 2000). Further tool information are reported in Table 3.

The tests were designed with the purpose of identifying the MUCT as a function of the feed per tooth. On each side of the 17-4 PH stainless steel workpiece, five cuts were performed moving the tool from the outer to the centre at a constant depth (a_p) of 200 μm. A cutting speed (v_c) of 40 m/min was kept constant for each cut. An actual tool diameter of 789 μm was measured by means of the BLUM laser measuring system mounted on the CNC machine.

Figure 1 illustrates the machining pattern and the load cell reference systems that was aligned to the KERN machine tool. Twenty micro channels were machined by

Table 3 Process parameters used in the micromilling process

Properties	Value
Nominal (μm)	800
Effective diameter (μm)	789 ± 2
Nominal cutting edge radius (μm)	5
Effective cutting edge radius (μm)	6.3
Helix angle (°)	20
Rake angle (°)	4
Material	Tungsten carbide
Material coating	Titanium nitride

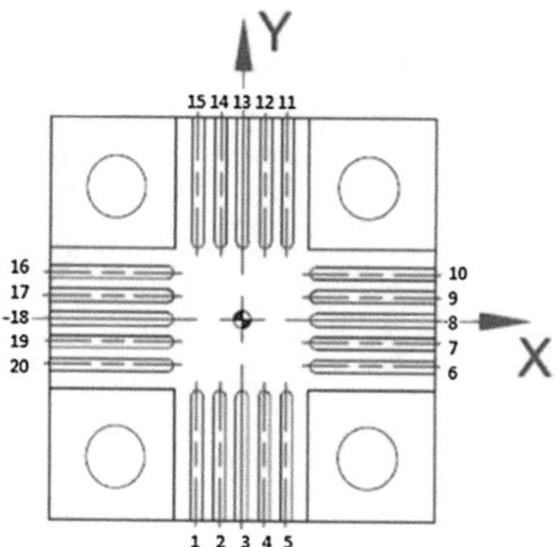

Fig. 1 Micro-slot pattern with the related reference systems

using twenty feed per tooth values ranging between 10 and 0.5 μm. The tool wear has been measured with a digital optical microscope and resulted negligible. Between two consecutive tests the tool has been properly cleaned to remove any stick material as confirmed by the optical microscope observations.

The force acquisition system allows to measure the cutting load component along each direction. The single components were subsequently composed to calculate the cutting force (F_c) through Eq. (1) directly in LabVIEW, the integrated development environment for the National Instruments graphic programming code.

$$F_C = \sqrt{(F_X)^2 + (F_Y)^2 + (F_Z)^2} \tag{1}$$

LabVIEW code was integrated by a Butterworth 20th order low-pass filter with a cut-off frequency of 1000 Hz. The tooth pass frequency corresponding to cutting speed and the tool can be calculated by Eq. (2).

$$f_{TP} = \frac{n}{60} * z \tag{2}$$

Considering the number of tool flutes ($z = 2$) and the spindle speed (n = 15,923 rpm), the tooth pass frequency is equal to 530 Hz and consequently it is lower than the cut-off frequency. The signal was filtered in order to identify the cutting force maximum peak on the flutes for each rotation. As shown in Fig. 2, the maximum peaks of the signal were not substantially altered.

2.3 *Evaluation of the Normalized Specific Force Fc*

The experiments allowed to investigate the regime transition as a function of the process f_z. The cutting force can not be directly used in this analysis and it must be normalized regarding the chip cross-section (S). Equation (3) shows the relation between chip section and feed per tooth:

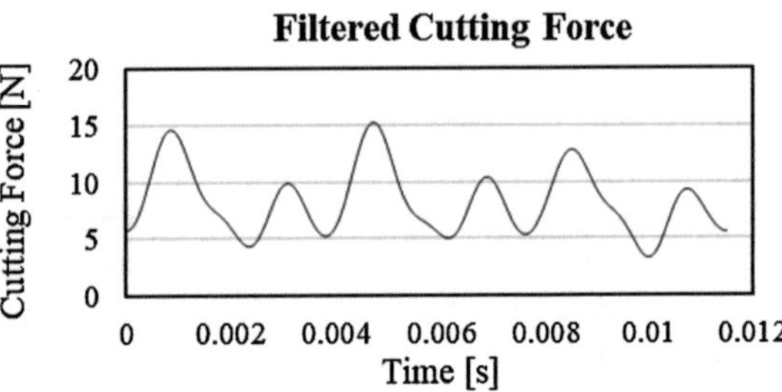

Fig. 2 Comparison between the original cutting force and the filtered cutting force

$$S = a_p * h \tag{3}$$

where a_p is the axial depth of cut while the chip thickness h can be calculated by Eq. (4)

$$h = f_z * \sin(\omega t) \tag{4}$$

In particular, the maximum cross section ($\omega t = \pi/2$) is expressed by the product between the axial depth of cut (a_p) and the feed per tooth (*fz*). The decrease of *fz* between two consecutive test determines a section (*S*) reduction which has a considerable effect on the cutting force value. To highlight the dependence of the cutting force in relation to the deformation mechanism, a specific cutting force must be calculated by Eq. (5):

$$F_{cn} = \frac{Fc}{S} \tag{5}$$

where Fc is the cutting force results from the combination of all the force components including the cutting edge component. The specific force allows to identify the MUCT. During ploughing regime, the workpiece material elasto-plastic deformation determines a load increment. The phenomenon is enhanced by the accumulation of uncut material against the cutting edge. When shearing regime is prevalent, the correct chip formation causes the specific loads decrease. A direct correlation between the regime transition and the specific cutting force must be identified without neglect the tool run-out effects. The tool run-out causes a difference between the effective chip thicknesses on each flute, determining an unbalanced load condition on the flutes. Considering two flutes, the maximum chip thickness for one flute (h_{Amax}) will be greater than the thickness for the other flute (h_{Bmax}). The asymmetric condition causes two cutting force peaks ($Fcmax_A$; $Fcmax_B$) which should be normalized by considering the effective thickness (see Fig. 2). Several tool run-out models should be utilized [8, 10, 11]. The simplest approach is based on the hypothesis of a direct relation between chip section and force peak. Equation (6) was implemented to calculate the effective chip thickness for tool flute A:

$$h_{Amax} = \frac{2*Fcmax_A}{Fcmax_B + Fcmax_A}*f_z \quad h_{Bmax} = 2f_z - h_{Amax} \tag{6}$$

where $\frac{Fcmax_B+Fcmax_A}{2}$ is the average force peaks between edge A and edge B (Fc_{av}). Supposing that the average undeformed chip thickness is equal to f_z, Eq. (6) derived from the proportion $h_{Amax} : Fcmax_A = Fc_{av} : f_z$. The signal of the Fc was considered in order to select a uniform portion corresponding to thirty tool rotations. For each spindle rotation the effective chip thickness h_{Amax} was calculated and subsequently an average value was obtained. The average value was finally utilized for the force normalization.

2.4 Roughness and Burrs Evaluation

The roughness was evaluated by means of a Mitutoyo SJ300 profilometer with a 2 μm tip. The width (W_B) and height (H_B) of each burr were measured for each inner and outer channel side. The burrs width was measured by using a Mitutoyo QuickScope optical coordinate measuring machine, while the burrs height was measured by using a Hirox RH-2000 optical microscope (Fig. 3).

The width and the height were combined supposing the absence of curvatures of the burrs to obtain a unique value of the length (L_B) of the burrs on each side

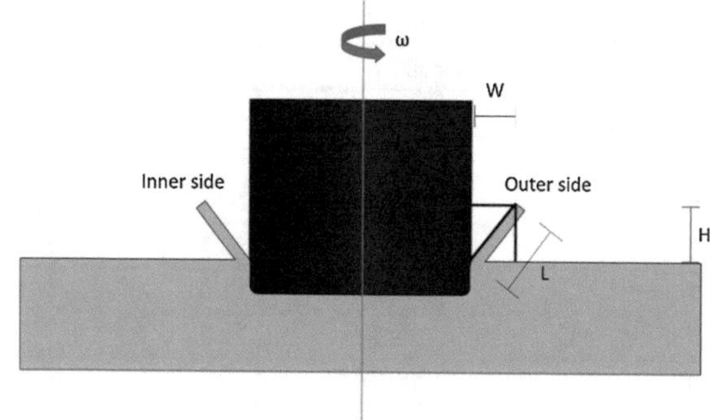

Fig. 3 Schematic representation of the geometry of the burrs

according to Eq. (7):

$$L_B = \sqrt{(W_B)^2 + (H_B)^2} \quad (7)$$

The collected data were reported as function of the feed per tooth in order to evaluate the dependence of roughness and burrs in relation to the feed rate.

3 Results and Discussion

In this section, the results related to the evaluation of the transition regime of deformation are reported. Tables 4 and 5 reports the values of the average specific cutting forces (Fc_n) calculated considering both the actual depth of cut (a_{peff}) and the force peaks ($Fcmax_A$, $Fcmax_B$).

Figure 4 shows the measured cutting force peaks for each flute. The difference between the force peaks can be related to the tool run-out. It is possible to observe that when the feed per tooth is higher than 2 μm, the tool run-out influence slightly increases as the feed per tooth decreases. In fact, it is evident that the difference between the force peak of the flute A and the force peak of the flute B increases as the feed rate decreases. When the feed per tooth is equal or lower than 2 μm, the difference between the cutting force peaks strongly increases as the feed per tooth decreases. This behaviour can be related to a ploughing condition involving flute B that causes an increment of the undeformed depth of cut for flute A. Consequently, the normalization performed on the resulting cutting force values allowed to investigate the cutting regime of the AM material by making the tool run out effects negligible.

Observing Fig. 5, the specific cutting force is not constant thorough the tests due to the presence of different deformation mechanisms. It shows an increase of the

Table 4 Results of the micromilling tests performed at different feed per tooth values ranging between 5.5 and 10 μm/tooth * rev

n	f_z (μm/t)	a_{peff} (μm)	$Fcmax_A$ (N)	$Fcmax_B$ (N)	Fc_n (N/mm^2)
1	10	205.4	18.7	14.1	8016.0
2	9.5	205.7	14.0	9.9	6127.6
3	9	201.3	13.2	9.1	6177.5
4	8.5	203.1	15.1	10.0	7266.7
5	8	203.3	10.4	8.6	5887.3
6	7.5	199.9	12.5	4.2	5595.4
7	7	198.1	11.0	1.9	4684.5
8	6.5	198.3	11.1	2.4	5251.8
9	6	200.7	11.4	2.5	5840.3
10	5.5	203.4	13.0	4.5	7857.1

Table 5 Results of the micromilling tests performed at different feed per tooth values ranging between 0.5 and 5 μm/tooth * rev

n	f_z (μm/t)	a_{peff} (μm)	$Fcmax_A$ (N)	$Fcmax_B$ (N)	Fc_n (N/mm^2)
11	5	200.5	12.2	4.1	8175.1
12	4.5	199.3	10.5	3.6	7960.0
13	4	198.9	11.7	2.5	9007.8
14	3.5	199.1	11.0	2.1	9515.1
15	3	198.7	10.9	2.1	10,994.4
16	2.5	209.8	10.6	1.6	11,766.8
17	2	206.3	15.2	3.1	22,277.7
18	1.5	200.8	15.6	4.3	33,253.7
19	1	205.8	17.3	3.0	49,514.9
20	0.5	202.6	17.8	1.9	97,726.5

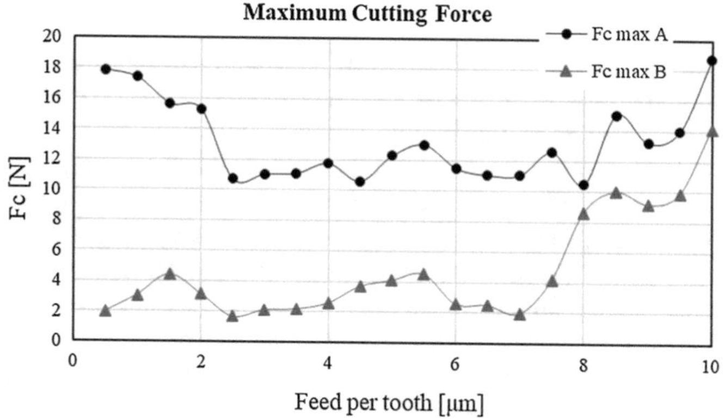

Fig. 4 The maximum cutting force value for each flute versus feed per tooth

normalized cutting force when the feed per tooth decreases from 2.5 to 2 μm/tooth. This trend allows to set the MUCT value to 2.5 μm/tooth. Considering the deformation mechanisms, region I corresponds to a ploughing dominated regime extended to a feed per tooth value of 2.5 μm/tooth according to literature [6], which identifies the MUCT value as the 20–40% of the cutting edge radius. The identified MUCT value is also in accordance with the behaviour observed in Fig. 4. On the other hand, region III is related to the shearing deformation regime characterized by an independence of the specific cutting forces from the feed per tooth values, as visible from the reported magnification of the specific cutting force. The region II can be identified as a transition zone between the two dominant deformations mechanism where the ploughing effects are progressively increasing as the feed per tooth decreases.

The roughness of the as built SLM parts was measured on three replica of the same samples and resulted in 12.85 ± 0.18 μm. The micro machining tests significantly

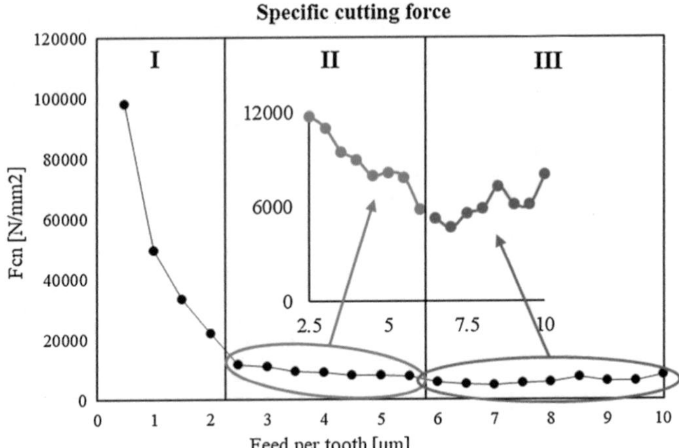

Fig. 5 Specific cutting forces versus feed per tooth

improved the surface quality as required by standard mechanical application. The effects of the occurring of the ploughing regime are visible on the evaluation of the roughness trend reported in Fig. 6.

As shown, high average roughness values are also related to the higher sensitivity to run out of the kinematic roughness at lower values of f_z. Moreover, the increased variability of the results is probably due to the incorrect chip formation mechanism during ploughing that is responsible of uncontrollable irregularities and accumulation of material thought the flute path. On the other hand, when the shearing regime is dominant, the mean roughness values depend on the feed per tooth values according to the typical trend of cutting tests. Moreover, the reduced variability of the data demonstrates the transition to a more regular deformation mechanism.

The burr length was considered to evaluate the influence of the ploughing regime on the feature quality (Fig. 7).

Fig. 6 Average roughness versus feed per tooth

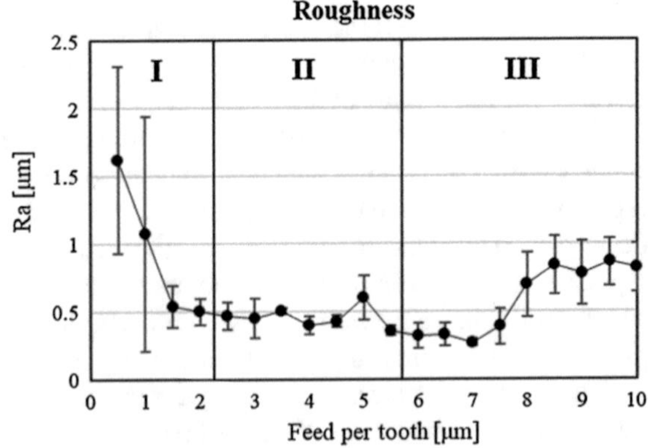

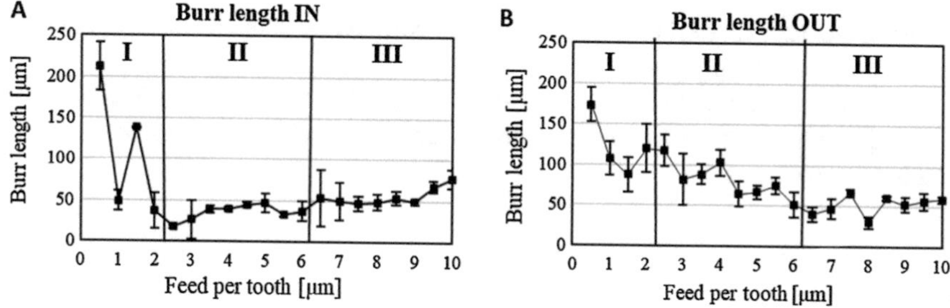

Fig. 7 Burr length on the inner side (**a**) and burr length on the outer side (**b**) versus feed per tooth

As expected, the length of the burrs on the outer side is higher compared to the length on the inner side.

The burr length on the inner side of the channels is strongly influenced by the occurring of the ploughing regime as highlighted from an increase of the average values. As the shearing regime is dominant, the length of the burrs begins to be dependent on the feed rate as visible from the trend of the data. On the other hand, the burr length on the outer side is more variable probably due to the instability of the ploughing regime and the presence of compressed AM material.

4 Conclusion

In this paper, the material removal behaviour of additive manufactured stainless steel was analysed. The specific cutting force resulting from the production of microchannels on SLM samples was evaluated as a function of the feed per tooth to identify the transition of the material from ploughing to shearing deformation regime. Moreover, the roughness and the burrs dimension of the machined workpiece were measured to investigate the effects of the deformation mechanism on the surface final finishing.

From the results it is possible to notice a ploughing dominated regime at low feed per tooth values followed by a transition to a shearing dominated regime at higher feed rates. In particular, the specific cutting forces related to the ploughing regime resulted ten times higher than the forces calculated when shearing regime is present. As expected, at the highest feed per tooth values the specific cutting force was found to be independent on the feed rate value. The results showed a behaviour transition of the material at the 30–35% of the effective tool edge radius. Thus, the study allowed to identify an optimal feed per tooth range related to reduced cutting forces with the aim of minimize the tool damage probability and consequently the tool wear rates.

Furthermore, the highest values of roughness were measured on the surface of the channels machined at low feed rates. The variability of the data showed a progressive decrease corresponding to a more dominant shearing regime. The dimension of the burrs calculated both on the inner and outer channel sides resulted higher at low feed per tooth values. Therefore, a high value of the feed rate is recommended to reduce the extension of the burrs and assure a correct mechanical coupling between the micro-sizes components.

A further development of this research will be based on the comparison between the additive manufactured and the conventionally produced components under machining operations.

References

1. Leary M (2017) Surface roughness optimisation for selective laser melting (SLM): accommodating relevant and irrelevant surfaces. In: Brandt M (ed) Laser additive manufacturing. Woodhead Publishing Series in Electronic and Optical Materials, pp 99–118
2. Galy C, Le Guen E, Lacoste E, Arvieu C (2018) Main defects observed in aluminum alloy parts produced by SLM: from causes to consequences. Addit Manuf 22:165–175
3. Kaynak Y, Tascioglu E (2018) Finishing machining-induced surface roughness, microhardness and XRD analysis of selective laser melted Inconel 718 alloy. Procedia CIRP 71:500–504
4. Camara MA, Rubio C, Abrao AM, Davim JP (2012) State of the art on micromilling of materials, a review. J Mater Sci Technol 28:673–685
5. Malekian M, Park S, Jun MBG (2009) Modeling of dynamic micro-milling cutting forces. Int J Mach Tools Manuf 49:586–598
6. Malekian M, Mostofa MG, Park SS, Jun MBG (2012) Modeling of minimum uncut chip thickness in micro machining of aluminum. J Mater Process Technol 212:553–559
7. Attanasio A (2017) Tool run-out measurement in micro milling. Micromachines 8:221
8. Attanasio A, Abeni A, Özel T, Ceretti E (2019) Finite element simulation of high speed micro milling in the presence of tool run-out with experimental validations. Int J Adv Manuf Technol 100(1–4):25–35
9. Attanasio A, Garbellini A, Ceretti E (2015) Force modelling in micromilling of channels. Int J Nano Manuf 11(5–6):275–296
10. Jin X, Altintas Y (2012) Prediction of micro-milling forces with finite element method. J Mater Process Technol 212:542–552
11. Zhang X, Ehmann KF, Yu T, Wang W (2016) Cutting forces in micro-end-milling processes. Int J Mach Tools Manuf 107:21–40

Integrating Machine Scheduling and Transportation Resource Allocation in a Job Shop: A Simulation Approach

Erica Pastore and Arianna Alfieri

Abstract In scheduling problems with fixed routing, usually the transportation of jobs among the machines is not considered (i.e., the transportation time between two stages is negligible, and the number of transportation resources is unlimited). However, in real contexts, this assumption can be unrealistic, especially when human supervision is needed for transportation, and hence not considering transportation can lead to low quality scheduling solutions. This paper considers a job shop in which transportation resources are limited and free to move among all the machines (no fixed routes). The aim is the integration of machine scheduling and transportation resource allocation, i.e., to decide for each machine the job sequence, and for each free transportation resource the routing. Due to the complexity of the problem, a Discrete Event Simulation approach is used to compare different scheduling and transportation resource allocation policies through scenario analysis.

Keywords Job shop · Machine scheduling · Transportation resource allocation · Discrete event simulation

1 Introduction

In manufacturing systems, different layouts are used to organize machines. Among them, the job shop allows to achieve the maximum flexibility in the production process. The job shop can handle a varying mix of products (that can be the result of the increasing variability in customer orders) to be produced in small batches and with different production cycles. Every product manufactured in a job shop has its

---

E. Pastore (✉) · A. Alfieri
Department of Management and Production Engineering, Politecnico di Torino, c.so Duca degli Abruzzi 24, 10129 Torino, Italy
e-mail: erica.pastore@polito.it

A. Alfieri
e-mail: arianna.alfieri@polito.it

E. Ceretti and T. Tolio (eds.), *Selected Topics in Manufacturing*, Lecture Notes in Mechanical Engineering,
https://doi.org/10.1007/978-3-030-57729-2_2

own operation sequence and, therefore, its own routing in the system. Moreover, since transports between machines are hardly automatized in the job shop, the position of the machines on the shop floor is chosen to limit the time wasted to move a batch from an operation to the following one. However, due to the variety of production cycles, many products might have to travel through the entire shop floor to fulfil all the required operations. For this reason, the job transportation between machines is a critical issue in the job shop management.

Although the relevance of the transportation issue, few research works have addressed the job shop scheduling with transportation resources, with respect to the amount of job shop scheduling literature in which transportation has been neglected (the standard assumption is that number of transportation resources is unlimited and the transportation time is negligible).

Even without transportation resources, the job shop scheduling problem is a very complex optimization problem and it belongs to the class of non-deterministic polynomial time (NP hard) problems [1]. Due to this reason, many of the approaches proposed in the literature are heuristic, as exact approaches (e.g., branch and bound or dynamic programming) can solve only small-scale problems. Just to cite few, not exhaustive, examples, the most common algorithms are genetic algorithms [2, 3], tabu search [4], and particle swarm [5, 6].

When transportation is included, the complexity of the scheduling problem increases, as the complete problem can be seen as the integration of two subproblems: a classical job shop scheduling problem and a vehicle routing problem. As previously mentioned, fewer papers have addressed it [7]. To illustrate some example, the flexible job shop scheduling problem in a cellular manufacturing environment has been considered in [8], including intercellular transportation times but omitting empty transportation times (i.e., the time the available transportation resource takes to arrive to the machine where there is a job needing to be moved). Also, the problem of simultaneous scheduling machines and AGVs in a flexible manufacturing system has been addressed in [9]. The automated guided vehicles do not have to return to the load/unload station after each delivery, and the problem is solved by an iterative procedure in which admissible time windows for the trip are constructed by solving the machine scheduling problem, which generates the completion times of each operation with a heuristic procedure. The flexible job shop scheduling problem with transportation constraints has been addressed in [10], where a set of identical transportation resources and empty transportation are considered, and a tabu search procedure is proposed for its solution. The classical job shop scheduling problem with transportation resources able to carry more than one task at a time has also been studied in [11].

From the examples discussed above, it clearly emerges that most of the literature focuses on the flexible job shop, in which each operation is not associated to a fixed machine but to a set of machines among which one has to be chosen. Although this problem can be harder to model and to solve than the job shop with fixed association among operations and machines (especially with exact solution approaches), the possibility to choose the machine can simplify the problem from the transportation resource standpoint. Moreover, to the authors' knowledge, no work focuses on the optimal schedule for the transportation resources, rather they are treated as additional time to be considered (thus, the objective function includes only the completion time).

In this paper, we consider the integrated job shop scheduling problem with transportation resources, in which the job shop is characterized by fixed routing and fixed association between operations and machines, and the transportation includes the empty transportation time. Differently from most of the papers in the literature, the objective function includes penalties for tardy jobs and transportation resource costs, with the aim of finding the optimal (from the economic standpoint) number of transportation resources, together with the optimal schedule of jobs on machines and on transporters.

The integrated job shop machine and transportation resource scheduling is modeled by a mixed-linear programming model. Due to the complexity of the problem, a simulation-optimization solution approach is proposed, and a case study from the textile industry is used to test its applicability in a real context.

The reminder of the paper is organized as follows. In Sect. 2 the problem is mathematically represented as a MILP model and its solution complexity is discussed. The simulation-optimization approach and its application to the case study are presented in Sect. 3. Section 4 concludes the paper discussing the limitations of the approach and future research directions.

2 Problem Description

As discussed in the previous section, the integrated job shop scheduling problem with transportation resources can be seen as a classical job shop, in which job operations have to be sequenced on machines, with the additional requirement of scheduling the transportation activities on the transporters. In the considered problem, the additional request of finding the optimal number of transportation resources is considered.

Specifically, let N be the jobs to process. Each job i has a set $\aleph_i$ of consecutive operations to be performed. To simplify the notation, it is assumed that operation j of job i is exactly the jth operation of the job in $\aleph_i$. The route of each job in the shop floor (i.e., the sequence of machines associated to the operations of the job) could be partially or entirely different from that of the other jobs.

Job i has a release date rd_i and a due date dd_i. If the job is not completed before dd_i, a tardiness penalty is paid. The processing time p_{ij} of operation j of job i is known and fixed, and so is the machine k_{ij} on which it has to be processed. Due to the limited number of transportation resources and to the not negligible transportation time between machines, for each job, each operation cannot start immediately after the end of the previous one, but the job has to be transported to the machine associated to the next operation. When a transporter becomes available, the next job to be transported must be decided considering both the completion time of jobs at their machines and the distance between the current position of the available transporter and the jobs waiting to be moved.

The objective is to select the appropriate number of transportation resources (also referred to as *transporters*, in the following), and to sequence all the jobs on the machines and on the transportation resources, in order to minimize the total cost of the tardiness and of the transportation resources.

Using the parameters and the variables summarized in Table 1, the integrated job shop machine and transportation resource scheduling problem can be modelled as follows.

$$\min \sum_{i=1}^{N} c_{TA} \cdot TA_i + \sum_{t=1}^{T} g_{TR}\delta_t \tag{1}$$

$$\text{s.t. } C_{i1} \geq rd_i + p_{i1} + d_{i1i2} \qquad \forall i \tag{2}$$

$$C_{ij} \geq C_{i(j-1)} + p_{ij} + d_{iji(j+1)} \qquad \forall\, i,\ j = 2, \ldots, n_i \tag{3}$$

Table 1 Parameters and decision variables of the mathematical model

Parameters	
c_{TA}	Unit cost of tardiness
g_{TR}	Unit cost of transportation resources
rd_i	Release date of job *i*
p_{ij}	Processing time of operation *j* of job *i*
k_{ij}	Machine associated to operation *j* of job *i*
n_i	Number of operations of job *i*
$d_{iji'j'}$	Distance (expressed in time units) between machines k_{ij} and $k_{i'j'}$
dd_i	Due date of job *i*
T	Upper bound on the number of transportation resources
M	Large positive number (the so-called big-M)
Decision variables	
$\mathbf{C_{ij}}$	Completion time of operation *j* of job *i*
$\mathbf{TC_i}$	Total completion time of job i
$\mathbf{WT_{i'j'ijt}}$	Time at which transporter t is available to transport job i at operation j if it had previously transported job i′ at operation j′
$\mathbf{TA_i}$	Tardiness of job i
$\boldsymbol{\beta_{iji'j'}}$	Binary variable equal to 1, if operation j of job i is scheduled before operation j′ of job i′; 0, otherwise
$\boldsymbol{\alpha_{ijt}}$	Binary variable equal to 1 if, if operation j of job i is assigned to transporter t; 0, otherwise
$\boldsymbol{\gamma_{iji'j't}}$	Binary variable equal to 1, if operation j of job i is assigned to transporter t before operation j′ of job i′; 0, otherwise
$\boldsymbol{\delta_t}$	Binary variable equal to 1 if transporter t is used; 0 otherwise

$$C_{ij} \geq WT_{i'j'ijt} + d_{iji(j+1)} \qquad \forall\, t, i \neq i', j = 1, \ldots, n_i, j' = 1, \ldots, n_{i'} \tag{4}$$

$$C_{ij} \geq C_{i'j'} + p_{ij} - M\beta_{iji'j'} \qquad \forall i \neq i', j \in \aleph_i, j' \in \aleph_{i'}, k_{ij} = k_{i'j'} \tag{5}$$

$$C_{i'j'} \geq C_{ij} + p_{i'j'} - \left(1 - \beta_{iji'j'}\right)M \qquad \forall i \neq i', j \in \aleph_i, j' \in \aleph_{i'}, k_{ij} = k_{i'j'} \tag{6}$$

$$WT_{i'j'ijt} \geq C_{i'j'} - p_{i'j'} + d_{i'j'ij} - M\left(2 - \alpha_{ijt} - \alpha_{i'j't} + \gamma_{iji'j't}\right) \\ \forall t, i \neq i', j = 1, \ldots, n_i, j' = 1, \ldots, n_{i'} \tag{7}$$

$$WT_{iji'j't} \geq C_{ij} - p_{ij} + d_{iji'j'} - M\left(3 - \alpha_{ijt} - \alpha_{i'j't} - \gamma_{iji'j't}\right) \\ \forall t, i \neq i', j = 1, \ldots, n_i, j' = 1, \ldots, n_{i'} \tag{8}$$

$$\sum_{t=1}^{T} \alpha_{ijt} = 1 \qquad \forall i, j = 1, \ldots, n_i \tag{9}$$

$$TC_i \geq C_{ij} \qquad \forall i, j = 1 \ldots n_i \tag{10}$$

$$TA_i \geq TC_i - dd_i \qquad \forall i \tag{11}$$

$$\gamma_{iji'j't} \leq \frac{\alpha_{ijt} + \alpha_{i'j't}}{2} \qquad \forall t, i \neq i', j = 1, \ldots, n_i, j' = 1, \ldots, n_{i'} \tag{12}$$

$$\delta_t \geq \frac{\sum_{i=1}^{N} \sum_{j=1}^{n_i} \alpha_{ijt}}{T} \qquad \forall t \tag{13}$$

$$C_{ij} \geq 0 \qquad \forall i, j = 1 \ldots n_i \tag{14}$$

$$TC_i \geq 0 \qquad \forall i \tag{15}$$

$$WT_{iji'j't} \geq 0 \qquad \forall t, i \neq i', j = 1, \ldots, n_i, j' = 1, \ldots, n_{i'} \tag{16}$$

$$TA_i \geq 0 \qquad \forall i \tag{17}$$

$$\beta_{iji'j'} \in \{0, 1\} \qquad \forall i, i', j = 1, \ldots, n_i, j' = 1, \ldots, n_{i'} \tag{18}$$

$$\alpha_{ijt} \in \{0, 1\} \qquad \forall t, i, j = 1, \ldots, n_i \tag{19}$$

$$\gamma_{iji'j't} \in \{0, 1\} \qquad \forall t,\ i,\ i',\ j = 1, \ldots, n_i,\ j' = 1, \ldots, n_{i'} \tag{20}$$

$$\delta_t \in \{0, 1\} \qquad \forall t \tag{21}$$

The objective function (1) minimizes the total tardiness penalty and the transportation resource cost. Constraints (2) state that the first operation of each job cannot be completed before its release date rd_i plus the first operation processing time and the time needed to move the job to the next machine. Constraints (3) and (4) ensure the precedence between consecutive operations of the same job. Specifically, constraints (3) represent the technological precedence while constraints (4) are needed as jobs are not always transferred to the next operation as soon as they are ready to be transported, as the transporters could be already busy in other transports. Constraints (5) and (6) guarantee that at most one part is processed by each machine at the same time. They are the classical *disjunctive constraints* and are used to sequence operations of different jobs requiring the same machine. Constraints (7) and (8) schedule the transporters and set their availability time. These constraints are only relevant when operation j of job i and operation j' of job i' are both assigned to the same transporter t (i.e., $\alpha_{ijt} = \alpha_{i'j't} = 1$). Constraints (9) assure that each transport is performed by a single transporter. Constraints (10) and (11) define the completion time of the last operation of job i and its tardiness, respectively. Constraints (12) link the binary variable used to assign each job to a transporter with the one used to schedule the transports assigned to every transporter. Constraints (13) are used to assess if a transporter is used. The number of used transportation resources is then given by $\sum_{t=1}^{T} \delta_t$. Finally, variable domains are set by constraints (14)–(21).

Due to the huge number of binary values and big-M constraints, the proposed model is hard to solve with standard approaches or commercial solvers (e.g., ILOG Cplex). In such a case, two alternatives are usually available: (1) to develop exact ad-hoc methods, mainly based on decomposition into sub-problems, which, however, can hardly address very large instances; (2) to use heuristic or meta-heuristic approaches, which can easily treat very large problems, but without any guarantee on the solution quality. In both cases, however, it is difficult to address the variability of processing times and of due dates (i.e., customers' orders).

To efficiently take the variability into account, in this paper the problem is solved by a simulation-based optimization procedure, implemented within a commercially available software. Specifically, Rockwell Arena simulation software is used to develop a Discrete Event Simulation model to replicate the job shop scheduling with transportation resources and to evaluate the performance of different scheduling and transportation policies with a fixed number of transporters; the commercial optimization tool OptQuest is used to vary the number of transporters to find the optimal one.

This approach is heuristic, as OptQuest adopts heuristic algorithms to solve the optimization problem, and the machine and transportation scheduling are both based on "rules" (e.g., maximum priority, minimum distance, etc.). However, it has the flexibility to easily address very different scenarios and, hence, to find bounds that could be further used in optimization approaches. For this reason, various operational problems are usually evaluated through simulation-optimization using commercially available softwares [12–14].

As the proposed approach is based on a simulation model, which is case-dependent, the case study will be presented before the discussion of the simulation-optimization model.

3 Case Study

As an example of integrated job shop machine and transportation scheduling, the finishing department of a textile company (that will remain anonymous for confidentiality reasons) has been considered.

The finishing department is the last phase, and one of the more complex departments, of the textile production. It includes very different processes made on many different product types, to assure that every manufacturing process can be properly completed. More than one thousand different items need to be finished in this department. They can be divided in two main families, worsted (used to make coats) and woollen fabrics (used to produce suits). The pieces of fabrics are often grouped in small lot sizes due to the large demand variety. Although all the final products are pieces of fabric, the sequence of the operations varies from item to item. For instance, at the beginning of the production process, worsted fabrics must be singed, to obtain an even surface by burning off projecting fibers, while woollen fabrics have to be carbonized, to remove vegetable fibers from wool in an acidic treatment. Moreover, within the same operation, a lot of differences can arise, as every piece of fabric can be washed and fulled in many ways (depending on the final aspect the product must have), thus resulting in very different processing times.

The fabric production cycles are often very long, as they include both wet and dry finishing operations, and a lot of transports are necessary to move every batch from a machine to the next one, especially when operations of the wet and dry finishing are done alternatively, and this usually takes a long time. The transportation issue becomes very critical in high demand periods, as the shop floor is almost 100% saturated and buffers are full. Currently, when the machine operator finishes processing a batch, she has to stop the machine and to transport the fabrics to the machine where the successive operation has to be done. Therefore, some machines risk being idle even if there are jobs to work, thus risking inefficiency in the system (e.g., lower service level). For this reason, the Company is evaluating the possibility of introducing some new operators to manage the fabric transportation, thus avoiding that the machine workers stop their processes.

In this context, the proposed simulation-optimization approach has been applied to quantify the impact of shop floor transportation on the global performance of the shop floor. Specifically, a simulation-based optimization model has been created to solve the cost optimization problem modeled in Sect. 2, and, hence, to identify the optimal number of transportation resources the Company should have.

3.1 The Simulation-Based Optimization Model

Simulation-based optimization procedures are usually exploited to solve complex optimization problems. They are traditionally composed of two detached modules that work iteratively until the optimal solution is found, or a defined stopping condition is met [15]. The optimization module gives as output a system configuration that is given as input to the simulator. The system performance of the proposed configuration is evaluated with the simulation, whose performance measures are given back to the optimization module [16].

In this paper, the simulator (implemented in Arena) evaluates the performance of different scheduling and transportation policies, given a fixed number of transporters (and fixed transportation and tardiness costs) as input. Referring to the mathematical model in Sect. 2, the simulator addresses all the constraints related to the scheduling and transportation dynamics, i.e., Eqs. (2)–(12). The optimization tool (OptQuest), instead, let the model vary the number of transportation resources. More generally, the optimization is used to define the scenarios to evaluate with the simulator, and to choose the optimal one. The objective function in Eq. (1) is evaluated, and various values of transportation resources are identified and given as input to the simulator. Figure 1 summarizes the simulation-optimization iterations, and the information given as input to the two modules.

The simulation module replicates the operations of the finishing department of the Company. As more than 1000 items are processed in the finishing department, to reduce the complexity of the simulation, they are grouped in 12 fabric categories, each one including items characterized by similar production cycles, and the finishing processes of these categories have been simulated ($i = \{1, \ldots, 12\}$). For each category, the due date distributions have been fitted from historical data (provided by the Company), and the same holds for the processing times of each operation. For each category, the set $\aleph_i$ of consecutive operations to be performed is given as input, and all the machines that perform the operations are modelled in the simulation environment. The set $\aleph_i$ contains from 8 to 20 operations for each fabric category. Each machine picks a job from its queue and processes it. To assure the minimization of the tardiness, the machines always pick the job with the closest due date. The machines process the fabrics in batches, whose size varies according to the specific fabric

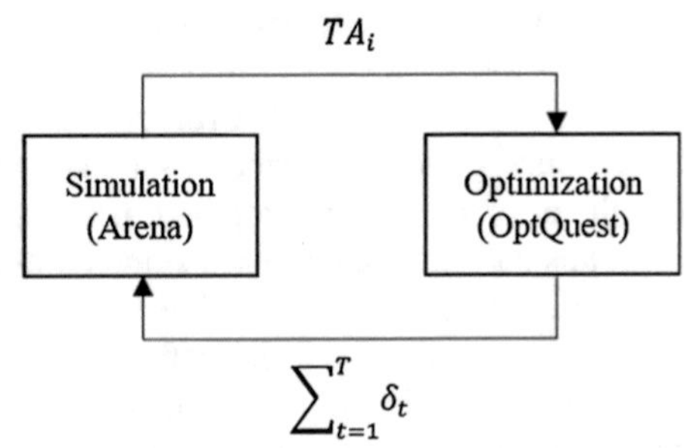

Fig. 1 Simulation-based optimization scheme

category. When a job finishes to be processed in a machine, it waits for an available transporter to be delivered to the next operation. When a transporter becomes idle, it moves to the closest machine with a job waiting to be moved. The total number of transporters per working shift is given as input from OptQuest.

Using as input the tardiness of the jobs given by the simulator, the optimization module finds the optimal number of transporters (i.e., the one that minimizes the objective function), and so on until no new solution is found by the optimization module.

3.2 *Experimental Design*

The simulation-optimization experiment has been designed as follows. The simulator, for each given number of transporters, performs 15 replicates of one year (i.e., the length of each replicate is 1 year of simulated time). This number of replicates has been chosen through the two-step method [15], and it leads to a reliable confidence interval of the throughput of the bottleneck machines (which is a critical performance measure for the considered production system).

To compute the objective function in (1), the unit cost of transporters and the unit cost of tardiness are needed (they are given as input to the optimization tool). The cost of the transporter g_{TR} has been estimated by the Company and it includes, in addition to the salary, all the training courses constantly done, the medical assurance provided by the Company, the subsidy for the meal in the canteen, the medical examinations each worker has to do periodically and their necessary equipment. The estimated transporter cost is not reported in the paper for confidentiality reasons. The cost of the tardiness, instead, is more complex to estimate. For some fabrics it is negligible, for others it might depend on the length of the delay, and sometimes tardiness might even cause the cancellation of the order and might contribute to the loss of a customer. For this reason, different levels of cost were considered. Thus, the daily unit cost of tardiness has been varied from 0 to 1000 €/(pcs * day) with a step of 5. Moreover, two different speed values have been considered for the transporter movements: 40 and 60 m/min.

3.3 *Numerical Results*

Figure 2 shows how the optimal number of transporters varies with different values of tardiness cost and transporter speed. With low speed (40 m/min), one transporter per working shift is the optimal solution if the daily unit tardiness is below 30 €/(pcs * day); two transporters per working shift are needed when the cost of tardiness increases above 30 €/(pcs * day). In this case, even increasing the tardiness penalty to unrealistic values, more than two transporters per working shift are never necessary. This is explained by the high cost of transportation resources together

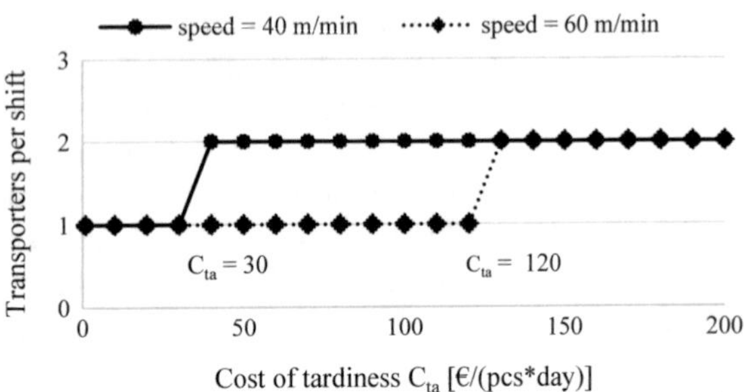

Fig. 2 Optimal number of transporters with varying cost of tardiness

with the small reduction in the total tardiness that an additional transporter would allow to reach. In other words, the saving in tardiness cost does not offset the cost of additional transporters.

Similar results are obtained for the case of transporters moving with high speed (60 m/min). In this case, two transporters are needed when the cost of tardiness is greater than 120 €/(pcs * day). The boundary unit tardiness cost increases with respect to the previous case because, as the transporters are faster in their movements, the total tardiness is smaller (being transported in a short time, the jobs will have a smaller completion time and hence a smaller tardiness, all the rest being equal) and one more transporter per working shift becomes necessary for larger costs of tardiness. Also in this case, however, three transporters are never necessary.

As no more than two transporters are required for each value of the tardiness cost, for readability reasons, Fig. 2 shows only values lower than 200 €/(pcs * day).

The above discussed system behavior can be more deeply analyzed by considering the low-speed case (as it is the most critical one). As reported in Table 2, with low speed, $C_{ta} = 40$ €/(pcs * day) and two transporters per working shift, 8.87% of the total fabrics produced in one year are delivered to customers with a delay, whereas the 9.33% of produced fabrics are late if only one transporter per working shift is used. With three transporters per working shift, no improvement can be appreciated, meaning that, when three transporters are available, the transporters are no longer the bottleneck of the process. Moreover, it appears that transporters can reduce the number of late jobs but not the average delay. This can be related to the naive priority rule approach in the management of machine and transporter queues; however, it gives an indication on the severity of the bottleneck and how it can move from transporters to machines, depending on the system conditions. This is also confirmed

Table 2 System performance with different transporters, speed = 40 m/min

Number of transporters	Percentage of fabrics with a delay (%)	Mean days of delay
0	9.37	13
1	9.33	13
2	8.87	13
3	8.87	13

by the comparison with the current situation with no transporter: the introduction of transporters is able to reduce the late jobs but not the average delay. Notice that, although the variation of the percentage of fabrics with delay is small, the savings of adding one more transporter can be relevant depending on the total yearly number of orders. For instance, in the case of 20,000 orders per year, switching from 0 to 1 transporter would decrease the number of fabrics with delay from 1874 to 1866 delayed fabrics. If the cost of tardiness is 40 €/day * pcs, then 4160 €/year are saved. If the tardiness is a relevant penalty (for instance, 125 €/pcs * day), then moving to 0–1 transporter would let the Company save 13,000 €/year. The yearly savings should be considered as part of a trade-off with the cost of hiring one more transporter and with the target customer service level the Company aims at achieving. The results for higher speed are similar, and for this reason, they are not reported in the paper.

The results discussed above depend on the numbers of jobs (i.e., of customer orders) that have to be processed and, since the fabrics produced and sold by the Company are affected by seasonality, the possibility of hiring a second seasonal transporter only for the months with larger demand must be evaluated. With larger demand, the machines are highly saturated, possibly causing some delivery delays. In this case, having more transportation resources available can assure a continuous and fast supply of every machine to prevent additional delay. To study this situation, the mean percentage of fabrics with delay and the mean delays have been considered separately for each month. The case of transporter speed = 40 m/min is reported in the following, but similar results hold for the case of higher speed.

As shown in Table 3, the months with larger production volumes correspond to the months with larger mean days of delay and percentage of delayed fabrics.

Table 3 Monthly system performance measures (speed = 40 m/min)

Month	One transporter		Two transporters	
	Mean days of delay	% of entities with a delay	Mean days of delay	% of entities with a delay
January	12	0.51	12	0.49
February	13	0.67	13	0.63
March	14	0.86	14	0.85
April	14	0.97	14	0.91
May	14	1.13	14	1.01
June	13	0.93	13	0.84
July	13	0.75	13	0.70
August	13	0.70	13	0.68
September	12	0.58	12	0.58
October	12	0.72	12	0.70
November	12	0.73	12	0.71
December	12	0.78	12	0.77
Total		9.33		8.87

During May, which is the month with the most critical delay (i.e., 1.13% of the annual produced fabrics are delayed in May), 0.12% of the annual produced fabrics on average are delivered on time by adding one transporter (i.e., an improvement of 10.5%). However, no improvement in the mean days of delay can be appreciated. The investment in hiring another transportation resource would be justified only if the cost of tardiness was very high compared to the cost of the transportation resource.

From the results, it clearly appears that the transportation is not always the bottleneck process of the finishing department. By analyzing the utilization of the machines, some of them can reach 100% utilization, especially in the peak-periods, and this is the main cause (in the current configuration of the finishing department and for the number of jobs causing these saturation levels) of the delays in the deliveries. If these bottlenecks were eliminated, by varying the number or speed of the machines, the schedule of every operation on each machine would surely change, and this change would possibly impact on the need for transportation resources of hiring more than one transporter per working shift.

4 Conclusions

Nowadays, customers demand a large variety of products in very short times, thus companies need to be flexible to respond as fast as possible to customers' orders. Managing thousands of different articles (characterized by different production cycles) and avoiding delays in product delivery to the customers (maintaining a high service level) are crucial issues for firms.

When the large variety of final products corresponds to a low production volume of each of them, to achieve the maximum flexibility, manufacturers usually design the shop floor as a job shop. However, due to the variety of production cycles, products travel all around the shop floor to fulfil their operations. As a consequence, managing together the job shop scheduling and the transportation among machines is a very relevant and critical issue.

This paper dealt with the integrated job shop machine and transportation resource scheduling problem in which also the optimal number of transporters to be included has to be chosen. A mathematical model that includes the job shop scheduling and the transportation routing was developed. By minimizing the cost of late deliveries (i.e., the cost of the total tardiness) and the cost of the transportation resources, the model optimizes both the scheduling of jobs on machines, the number of transportation resources needed in the shop floor per working shift, and the scheduling of jobs on transporters.

The solution procedure for the resulting mathematical model is highly complex and time consuming, thus a simulation-based optimization procedure was developed. The simulation module is used to evaluate the performance of naive scheduling and transportation policies, given a fixed number of transporters. The optimization module finds the optimal number of transporters, given the unit tardiness cost and the unit transportation resource cost. The two modules iterate exchanging the respective output until no new solution is found.

The simulation-based optimization procedure was tested in a real case study of the finishing department of a textile company. More than one thousand different products need to be finished in the department, each of them with a specific sequence of operations, performed in tens of different machines within the shop floor. Currently, the machine operators move the jobs from one machine to the other, causing inefficiency and large delays. The model developed in the paper and the simulation-based optimization solution procedure were used to find the optimal number of transporters per working shift to be hired by the Company.

The solution procedure was implemented using Rockwell Arena and OptQuest softwares, and various scenarios of tardiness cost and transporter speed were evaluated, whereas the cost of the transporter was considered as fixed. Results showed that two transporters are enough to minimize the delays related to the transportation. In fact, when the number of transportation resources is larger than 2, some of the machines become the bottleneck of the department, which are 100% saturated independently from the number of transporters in the job shop.

Although the interesting results, which highlight the importance of correctly and efficiently managing transportation in complex shop floors as job shops, due to the interaction between transportation and machine utilization, the simulation-optimization approach developed in the paper is a heuristic approach, and, hence, give no assurance about the quality of the solution. Future research will address the development of an exact algorithm able to solve the job shop scheduling with transportation model, which has been formalized in the paper. Due to the complexity of the complete model, the exact algorithm should exploit some properties of the system. For instance, as the problem includes both a job shop scheduling and transportation issues, approaches based on a decomposition of the problem in these two aspects could be considered. In this case, attention must be paid to the coordination between the two sub-problems. Possible schemes are Bender [17] or Dantzig-Wolfe [18] decompositions.

References

1. Garey EL, Johnson DS, Sethi R (1976) The complexity of flow shop and job shop scheduling. Math Oper Res 1:117–129
2. Tamilarasi A, Anantha KT (2010) An enhanced genetic algorithm with simulated annealing for job-shop scheduling. Int J Eng Sci Technol 2:144–151
3. Wang YM, Xiao NF, Yin HL, Hu EL, Zhao CG, Jiang YR (2008) A two-stage genetic algorithm for large size job shop scheduling problems. Int J Adv Manuf Technol 39:813–820
4. Eswaramurthy VP, Tamilarasi A (2009) Hybridizing tabu search with ant colony optimization for solving job shop scheduling problems. Int J Adv Manuf Technol 40:1004–1015
5. Jamili A (2018) Job shop scheduling with consideration of floating breaking times under uncertainty. Eng Appl Artif Intell 78:28–36
6. Li J-Q, Pan Y-X (2013) A hybrid discrete particle swarm optimization algorithm for solving fuzzy job shop scheduling problem. Int J Adv Manuf Technol 66:583–596

7. Nouri HE, Driss OB, Ghédira K (2016) A classification schema for the job shop scheduling problem with transportation resources: state-of-the-art review. In: Silhavy R, Senkerik R, Oplatkova Z, Silhavy P, Prokopova Z (eds) Artificial intelligence perspectives in intelligent systems. Advances in intelligent systems and computing, vol 464. Springer, Cham, pp 1–11
8. Deliktas D, Torkul O, Ustun O (2017) A flexible job shop cell scheduling with sequence-dependent family setup times and intercellular transportation times using conic scalarization method. Int Trans Oper Res
9. Ulusoy B (1995) A time window approach to simultaneous scheduling of machines and material handling system in an FMS. Oper Res 43:1058–1070
10. Zhang Q, Manier H, Manier MA (2012) A genetic algorithm with tabu search procedure for flexible job shop scheduling with transportation constraints and bounded processing times. Comput Oper Res 39:1713–1723
11. El Khoukhi F, Lamoudan T, Boukachour J, El HilaliAlaoui A (2011) Ant colony algorithm for just-in-time job shop scheduling with transportation times and multirobots. ISRN Appl Math 2011:1–19
12. Netto JF, Botter RC (2013) Simulation model for container fleet sizing on dedicated route. In: Proceedings of the 2013 winter simulation conference
13. Pawlewski P, Hoffa P (2014) Optimization of cross-docking terminal using flexsim/optquest—case study. In: Proceedings of the 2014 winter simulation conference
14. Waldemar M (2014) Cost optimization in manufacturing system with unidirectional AGVs. Appl Mech Mater 555:822–828
15. Kleijnen JP (2008) Design and analysis of simulation experiments. In: International series in operations research and management science, vol 111. Springer
16. Law AM (2015) Simulation modelling and analysis, 5th edn. McGraw Hill, Boston
17. Geoffrion AM (1972) Generalized benders decomposition. J Optim Theory Appl 10(4):237–260
18. Vanderbeck F (2000) On Dantzig-Wolfe decomposition in integer programming and ways to perform branching in a branch-and-price algorithm. Oper Res 48(1):111–128

Toolpath Optimization for 3-Axis Milling of Thin-Wall Components

Niccolò Grossi, Lorenzo Morelli, and Antonio Scippa

Abstract Milling of thin-wall components often entails significant workpiece static deflections, which make manufacturers use conservative cutting parameters along the toolpath to meet the tolerance required. This paper presents a technique to define the 3-axis toolpath that maximizes cutting parameters, without compromising the accuracy of the component. This goal is achieved by coupling a FE model of the workpiece, updated to include material removal mechanism, to a mechanistic model of the cutting forces. The algorithm follows the milling cycle in the reverse order: starts from the finished part, computes the maximum allowable radial depth of cut, and adding material accordingly, generates the toolpath until the stock is build. The proposed technique has been experimentally validated through comparisons between milling tests and numerical results, both traditional and optimized toolpaths have been tested to assess accuracy, benefits and limitations of the method.

Keywords Toolpath · Milling · Thin-wall workpiece

1 Introduction

Milling is one of the most used technology in the mechanical industry thanks to its versatility and its accuracy. Unfortunately, the productivity of the process for flexible thin-wall parts, typical of the aerospace and energy sectors, is still limited by accuracy issues. Indeed, thin-wall components low rigidity entails errors on the surface caused

N. Grossi (✉) · L. Morelli · A. Scippa
Department of Industrial Engineering, University of Firenze, Via di Santa Marta 3, 50139 Firenze, Italy
e-mail: niccolo.grossi@unifi.it

L. Morelli
e-mail: lorenzo.morelli@unifi.it

A. Scippa
e-mail: antonio.scippa@unifi.it

E. Ceretti and T. Tolio (eds.), *Selected Topics in Manufacturing*,
Lecture Notes in Mechanical Engineering,
https://doi.org/10.1007/978-3-030-57729-2_3

by static deflection [1] and vibrations [2]. Focusing on dimensional tolerance (i.e., workpiece geometry errors), the key factor impacting on the accuracy is to be found in the workpiece static deflection caused by the cutting forces [3]. Since in-process methods to tackle this issue [4, 5] are expensive, time-consuming and difficult to be applied, virtual predictive techniques are the preference choice to build an integrated approach. These methods compute workpiece displacements based on cutting forces and workpiece compliance.

Different strategies have been proposed in literature to both handle the workpiece deflection and ensure the required tolerance: the ones focused on error compensation through a modified toolpath [6, 7] and the ones focused on the optimization of the process parameters (e.g., radial depth of cut) [8, 9].

Rao and Rao [6] developed an error compensation technique, which employs an analytical/numerical model to predict the part deflection that is adopted to compensate the toolpath. Ratchev et al. [7], instead, starts from an analytical force model integrated with a Finite Element (FE) model to estimate the workpiece deflection which is then used to offset the toolpath. The main disadvantage of compensation methods in 3-axis milling operations is that tool helix angle produces a variable surface error along the axial depth of cut direction [10], hence hardly to be compensated by a simple translation of the toolpath.

On the other hand, Wang et al. [8] and Koike et al. [9] proposed a parameter selection system to deal with workpiece deflections. Essentially, constant axial and radial depths of cut, chosen on maximum allowable deflection basis, are used to build blocks on the finished workpiece. Each one of these blocks represents a volume of the material that should be removed to obtain the final shape of the component. Sequence of cutting is defined starting from the finished part adding blocks according to the stiffness of the workpiece, till the whole stock is created. The main disadvantage of the method relies on the cutting parameters estimation that considers the predicted deflection error without compensating it. Moreover, since it is based on constant cutting parameters, axial and radial depths of cut are constrained by the most flexible point of the component, leading to low productivity, especially in the stiffest areas of the workpiece.

This paper presents a technique, for 3-axis milling operations, that generates an optimized toolpath ensuring the geometric tolerance on a thin-wall component. This strategy combines the benefits of the error compensation methods and parameters selection approaches and it considers the surface error variation along the axial depth of cut. Indeed, the method compensates the mean deflection which causes the geometric error and selects the process parameters to ensure that the deviation of the deflection, which cannot be compensated, satisfies the tolerance. In detail, the proposed method couples the actual workpiece stiffness during the material removal process, computed using a FE model based on 2D shell elements (Sect. 2.1), with the error prediction along the axial depth of cut obtained from the approach presented in [10] (Sect. 2.3). The algorithm starts from the final geometry, finding the most suitable point (i.e., end of the toolpath). Then it proceeds to the next points, adding material (according to the computed radial depth of cut), following the milling cycle in the reverse order, as in [8], but considering a variable radial depth of cut (Sect. 2.2).

Elaborating the order of the identified points and related radial depth of cut, optimized toolpath is generated (Sect. 2.4). The proposed technique has been experimentally validated on the 3-axis milling of a blade (NACA 0005 airfoil blade) (Sect. 3).

2 Proposed Approach

The proposed approach aims at computing the most suitable toolpath for 3-axis milling of thin-wall components to meet the target tolerance. The main features are: (i) best cutting sequence identification, (ii) radial depth of cut identification and (iii) compensation of the predicted surface error on the toolpath. To achieve the goals, the method, schematized in Fig. 1, is composed by several blocks summarized in four steps:

- Numerical model and workpiece stiffness prediction;
- Selection of layer and path (cutting sequence);
- Error along the axial depth of cut;

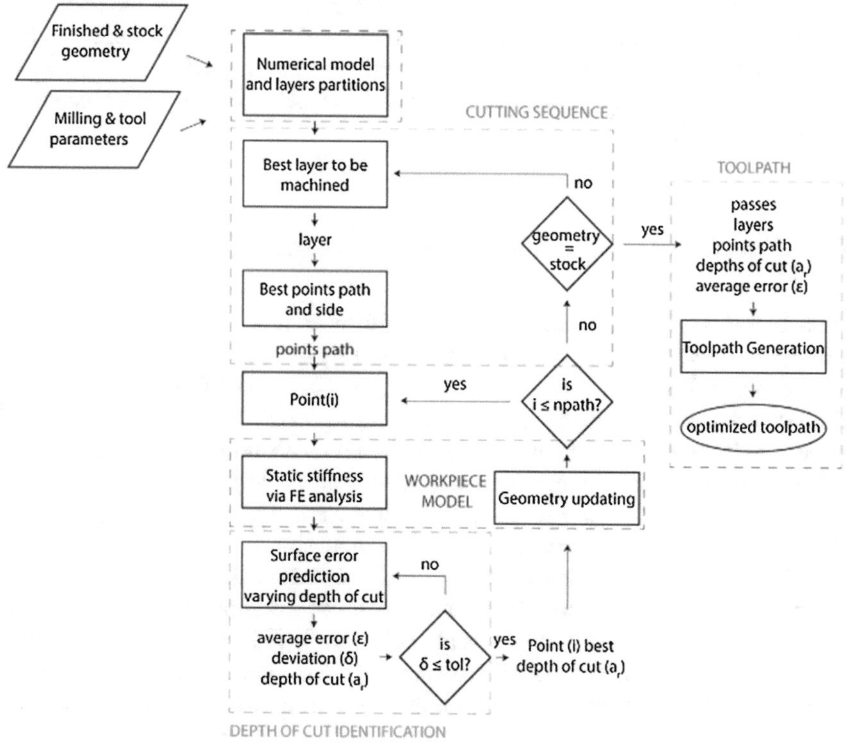

Fig. 1 Proposed approach scheme

- Toolpath generation.

Using the inputs (process parameters and tool data), the method reconstructs the best toolpath in a reverse order: starting from the finished part till the stock geometry is reached. First, the numerical model of the finished part is created by discretizing its geometry (i.e., FE model). Based on the axial depth of cut, the different layers to be machined are identified. For each layer, the machining allowance is computed as the difference on thickness between finished and stock geometry. The algorithm starts the loop by selecting the best layer among the ones available (i.e., the ones in which the machining allowance is not zero). From the selected layer on the finished part, the best path (cutting sequence) is then identified (i.e., the order of points). On each point in sequence, the radial depth of cut computation is performed. Radial depth of cut is selected as the highest value that allows to meet the target tolerance (input of the algorithm). Surface error is estimated by a dedicated algorithm that exploits cutting force prediction and static deformation of the workpiece in the cutting point, computed using FE analysis on the model. Once the radial depth of cut is identified, workpiece allowance and numerical model are updated. Radial depth of cut is computed for all the points along the path and the procedure is repeated till the stock geometry is reached. The sequence of layers, paths, radial depths of cut and surface errors are then used to build the toolpath, starting from the end of the computed cycle till the first step (i.e., finished part) is obtained. The algorithm, developed in Matlab, automatically arranges the FE analysis, performed using MSC Nastran. In the following sub-sections, the different blocks of the proposed technique are examined in depth.

2.1 Numerical Model and Workpiece Stiffness Prediction

Predicting workpiece stiffness is not a straightforward process, since the workpiece changes its geometry and hence its behavior during the machining cycle. Different strategies have been proposed in literature [5, 11, 12] most of them implies the use of Finite Element (FE) models, a convenient way to enable a virtual identification of workpiece behavior at different points and different machining steps. Ratchev et al. [11] developed an error compensation strategy for single pass peripheral milling based on FE method, while Budak [12] studied static displacements of cantilever plates milling with slender end-mills. Other authors simulate thin-wall components behavior for dynamics prediction purpose: Bolsunovskiy et al. [13] propose a method to compute the best spindle speeds to reduce the forced vibrations based on FE models of the thin-wall components, Tuysuz et al. [14] applied a reduced modeling technique on a full FE model of a thin-wall structure to predict chatter. All these works are based on 3D solid elements FE models, that for thin-wall components implies the use of small dimension elements, leading to high computational cost. This could be unacceptable when stiffness prediction should be performed several times to study workpiece deflection along the toolpath. In this work, the use of 2D shell elements

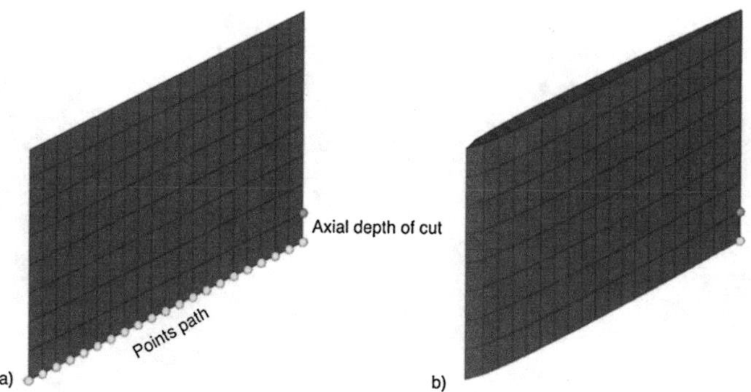

Fig. 2 Shell elements **a** 2D representation and characteristic nodes, **b** 3D representation

is proposed with a twofold advantage: (i) reducing computational efforts, (ii) easing both the automatic generation and updating of the thin-wall structure. Shell elements are suitable for thin-wall components and ensure the modeling of complex structure by using the mid-surface and variable thickness. In this work, mesh size on axial direction is selected equal to the axial depth of cut, while the mesh size on the other direction is selected to guarantee an accurate workpiece geometry discretization and behavior analysis.

The nodes on the mid line are the ones in which the optimal radial depth of cut is evaluated. On the axial direction two nodes are located limiting the axial depth of cut, as exemplified in Fig. 2. As previously mentioned, during the algorithm the mesh is automatically updated, allowing a fast generation of the toolpath. This is possible thanks to few operations: (i) node id are numbered in a specific order to be used by the algorithm to compile the analysis deck and launch the FE solver (e.g., Nastran), (ii) shell elements thickness will be based on the values given to the nodes (CQUAD4 card in Nastran [15]). This allows the procedure to automatically update the thickness of the nodes at each step by adding the computed radial depth of cut. In Fig. 3 an example of mesh updating procedure is presented. The algorithm starts with the mesh of the finished part (Fig. 3a), at the first point, radial depth of cut is computed, and thickness of the part is updated on the shell element (Fig. 3b), this procedure continues till all the nodes are analyzed and mesh updated (Fig. 3c). On the updated FE model of the workpiece at each step, static stiffness on the analyzed points is evaluated by performing a linear static analysis (SOL 101 in Nastran).

2.2 *Selection of Layer and Points Path*

The first step of the method is the identification of the cutting sequence. The idea is to build the toolpath, analyzing the cycle in the reverse order. The algorithm starts from the finished part and, at each step, finds the point where workpiece displacement

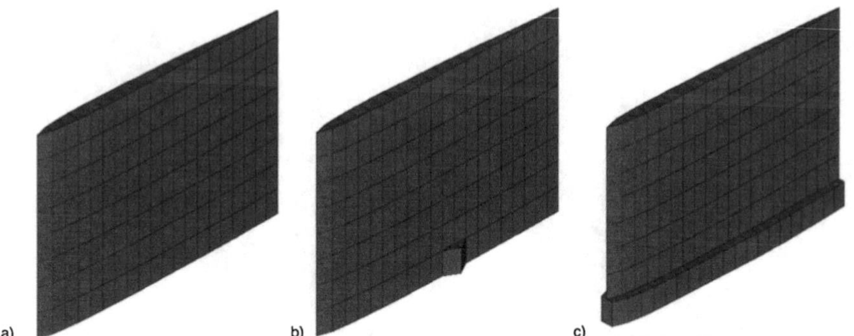

Fig. 3 Thickness updating **a** finished part, **b** first point updating, **c** full pass updating

due to the cutting forces is minimum. In that point, material is added (machining allowance is updated) according to the computed depth of cut, which is the maximum allowable value to meet the tolerance (Sect. 2.3). The algorithm stops when all the machining allowance is added to the final workpiece shape: i.e., stock geometry is achieved. Finally, the sequence is reversed to obtain the removal process. Cutting sequence in 3-axis milling is divided in: the layer to be machined (Z-axis) and the path on X–Y plane (points path), represented by the two nested loops of the algorithm (Fig. 1): (i) best layer selection; (ii) points path evaluation.

The first defines the axial position of the tool, while the second evaluates X–Y toolpath on the specific layer, analyzing all the points on the path. A new selection of the layer is repeated after that the points path evaluation procedure is completed, till the entire machining cycle is reconstructed. In general, layer selection should consider both machining constraints and flexibility of the components at several points, analyzing all the layers still to be machined (since the procedure is in the reverse order, a layer cannot be selected when it has already reached the stock dimension). This approach could not be trivial, since complex components and boundary conditions could benefit from a specific cutting sequence that enables the machining of the part in different zones in different conditions (i.e., time instant). In this paper, layer selection was implemented as simple as possible, considering only the flexibility of the component on one node at the same X–Y position and at the different heights (i.e., layers), performing a static analysis and choosing the stiffest layer. This approach is suitable for the investigated case of a cantilever plate and results in machining entirely each layer before going to the next from the top to the bottom. On the other hand, a more complex procedure that considers technological precedence and stiffness of the part in different points, not yet developed, is required in case of a more general application.

Once the layer is selected, points path within the layer is identified. For all the nodes of the layer, static stiffness is evaluated and the stiffest one is selected as the first node (i.e., the end of the path). In this work, the method is developed for open thin-wall components that needs to be machined on both sides. Therefore, the points path will pass through the same node twice. A procedure as the one proposed by

Wang et al. [8] would require the computation of the workpiece static stiffness in all the machinable points at every step of the machining cycle, leading to a very high computational cost. On the contrary, since a continuous machining cycle is preferable (as pointed out also in [8]), it is only required to define the first point and which side machines first. The side is selected based on the operation type (up or down milling) in order to reach the most flexible node (i.e., lowest static stiffness) on both sides when most of the material is still to be machined. Indeed, material on the part increases the local stiffness of the component, decreasing its deflection. Once the points path is defined, nodes in sequence will be analyzed to compute the maximum radial depth of cut allowable to meet the target tolerance, as explained in the next sub-section.

2.3 Error Along the Axial Depth of Cut

In peripheral milling, machined surface is generated at the instant in which tool passes over it (cutting edge perpendicular to the surface). Due to the helical nature of cutter, the axial location of surface generation points changes continuously with cutter rotation and the surface is created at different time instants. Since cutting forces are not constant over time, surface points along the tool axis are generated under the effect of different cutting forces. Therefore, even considering negligible the difference in terms of deflection between tool and workpiece along the tool axis, the surface error caused by the deflection of both the workpiece and the tool will change along the axial depth of cut and cannot be simply compensated by translating the original toolpath. In this work, surface error along the axial depth of cut is predicted by the formulations presented by Desai and Rao [10], extended to workpiece deflection, and a tailored compensation strategy was adopted. The surface error distribution depends on several factors, such as milling strategy (up or down milling), tool geometry (number of flutes and helix angle) and process parameters (axial and radial depth of cut). Starting from these input values, the three characteristic angles can be computed, α_{en} (radial engagement angle), α_{sw} (axial engagement angle) and φ_p (pitch angle) and used in the formulations to define the error shape. Error magnitude will depend on cutting forces and tool-workpiece stiffness. In this work, cutting force prediction is based on mechanistic force model adopting the following equations:

$$dF_t = K_t hdb; \quad dF_r = K_r hdb; \quad dF_a = K_a hdb \tag{1}$$

where F_t, F_r, F_a are the tangential, radial and axial components of cutting force respectively, h is the uncut chip thickness and *db* is the chip width. Each of the force component is described by one coefficient related to material shearing and proportional to the chip thickness (K_t, K_r, K_a). As far as stiffness is concerned, tool stiffness will be measured and stored in the algorithm as input, while workpiece

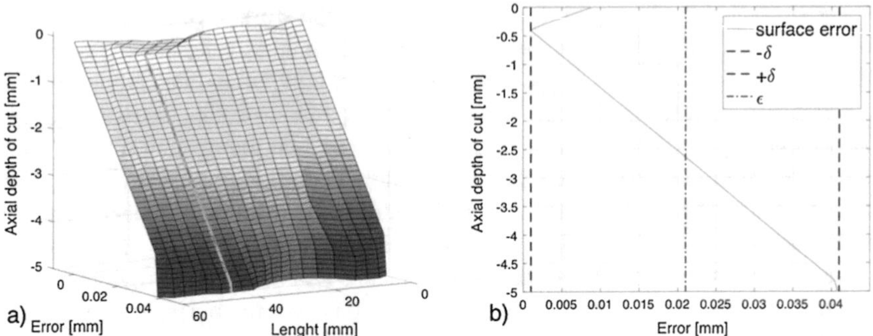

Fig. 4 Example of surface error **a** surface error prediction, **b** two error components

stiffness is computed using the FE model presented in the previous section. Using these data surface error distribution can be predicted and used to define toolpath.

In this method, two error components are distinguished: an average error (ε) and a deviation error (δ), as shown in the example in Fig. 4. The first (ε) will be compensated by offsetting the original toolpath, while the second (δ), that cannot be compensated in 3-axis, will be used in the algorithm to compute the most suitable radial depth of cut. Indeed, the value of deviation (δ) will change with the radial depth of cut and the maximum allowable value will be calculated in order to meet the target tolerance.

2.4 Toolpath Generation

The proposed algorithm (described in the previous sections), automatically selects: (i) the Z-axis position (i.e., layer), (ii) points path and (iii) radial depths of cut of the different passes to build the stock, starting from the finished part.

After collecting these data, toolpath is generated just reversing the passes and the points path and offsetting the stock geometry of Δ:

$$\Delta^i = \sum a_r^f + a_r^i + \varepsilon \tag{2}$$

where a_r^f is the radial depth of cut of the following passes on the same node, a_r^i is the radial depth of cut computed for the specific pass and point, ε is the average error of the surface to be compensated. The sign of the offset depends on the side of the workpiece, sign of the error (ε) depends on the operations (down or up-milling).

3 Experimental Validation

The proposed approach has been experimentally validated on a 3-axis milling of a thin wall component, as case study. A NACA 0005 airfoil profile (60 × 40 mm), made of aluminum (6082-T4) has been machined on a DMU 75 machine tool, using a 12 mm diameter, four-fluted endmill (Garant 202552), starting from a stock with 6 mm of thickness and 60 mm overhang out of the clamp. To apply the proposed strategy, cutting force coefficients (Eq. 1) have been computed by using a mechanistic approach, based on the average cutting forces acquired on a specimen of the same material in slotting at 5 different feed per tooth (0.05–0.1–0.15–0.2–0.25 mm) and 2 depths of cut (0.5–1.0 mm) using a Kitsler 9257A table dynamometer (Fig. 5a). Experiments were replicated 3 times to improve the reliability and the resulting coefficients, including Confidential Interval CI (95%), are reported in Table 1. The toolpaths have been used to machine the test case on the CNC machine using the compensation of the cutter radius (G41), measured using an on-machine laser measuring system (BLUM Micro Compact NT 87) (Fig. 5d). Surfaces have been acquired using the on-machine probe (RENISHAW PowerProbe 60) (Fig. 5c). Tool static stiffness has been identified analyzing displacement/force Frequency Response Function, acquired using laser displacement sensor (Keyence LK-H085) and impact hammer (PCB 086C03) (Fig. 5b). The target tolerance of the part was set to ±0.02 mm on a single side (i.e., ±0.04 mm on the thickness of blade). These data along with tool and cutting parameters are summarized in Table 1.

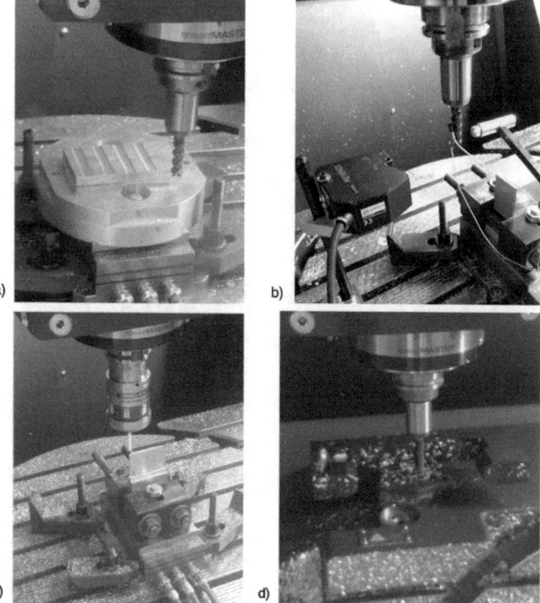

Fig. 5 Machine set-up **a** cutting force coefficient tests, **b** tool static stiffness measure, **c** machining error acquisition, **d** tool radius measure

Table 1 Tool and cutting parameters

Tooling			Tool static stiffness			Cutting parameters				Cutting force coefficients	
Helix angle (°)	D (mm)	z	k_{tx} (N/mm)	k_{ty} (N/mm)	Tolerance (mm)	Strategy	ap (mm)	v_c (m/min)	f_z (mm)	K_t (N/mm^2) CI (95%)	K_r (N/mm^2) CI (95%)
45	12	4	4110	5710	±0.02	Down milling	5	200	0.1	752.9 ± 22.0	200.5 ± 12.5

3.1 Toolpath

Both a traditional toolpath and the optimized toolpath have been tested, using the same cutting parameters. Traditional toolpath consisted in leaving a small machining allowance (0.5 mm) over the finished part to be removed with the final passes. Finished and stock geometries were input to the developed algorithm and the procedure starts building the FE models (Fig. 6a–c). The FE model of the part is composed by 160 3×5 mm shell elements (CQUAD4 in Nastran), while the substrate (part not to be machined) was modelled via solid elements. The following mechanical characteristics were considered for the aluminum: elastic modulus 72.5 GPa, density 2680 kg/m^3, Poisson's ratio 0.34. Following the steps summarized in the previous sections the method reconstructed the optimized toolpath. Depths of cut and errors are computed by FE simulation on the nodes of the model, however generating the part program starting from FEM nodes by linear interpolation will imply a low resolution of the part or a very high computational cost (i.e., if the number of nodes is increased). Therefore, the part program is generated by linear interpolation of the toolpath on a new denser discretization of the geometry (50 times the number of

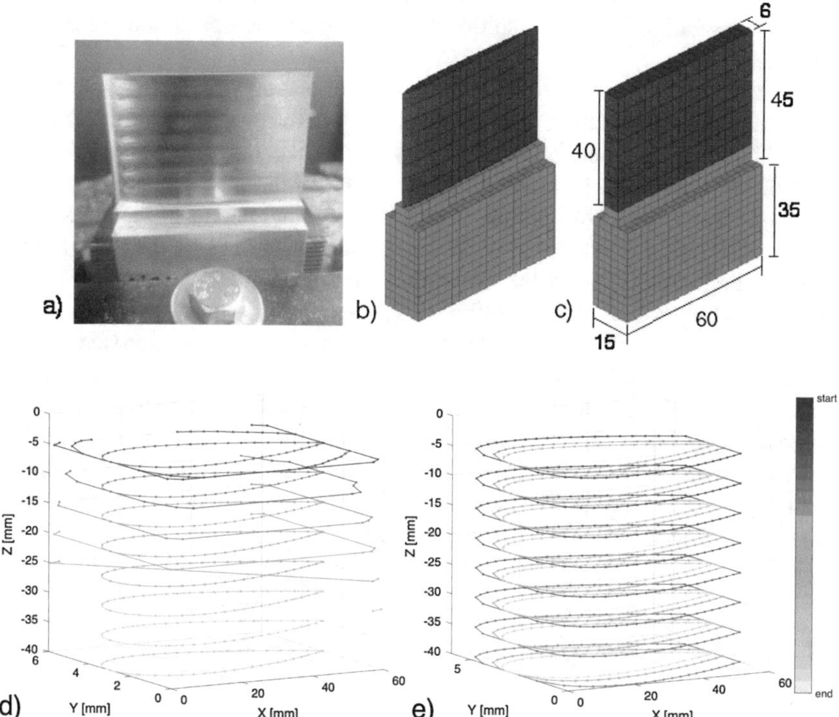

Fig. 6 Case study **a** physical, **b** finished FE model, **c** stock FE model with dimensions in mm, **d** optimized toolpath, **e** traditional toolpath

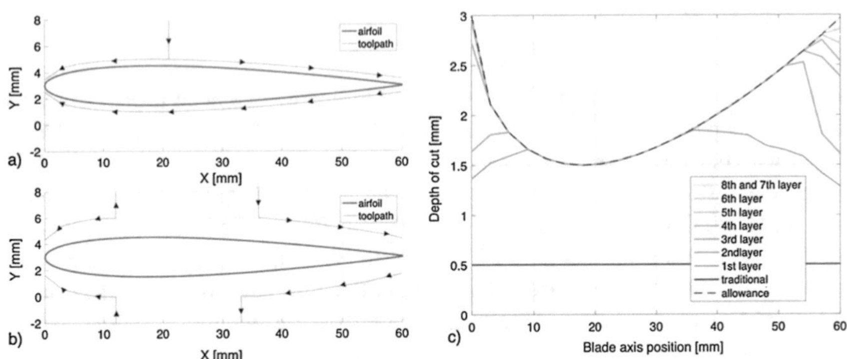

Fig. 7 **a** Example of traditional pass, **b** optimized pass, **c** depth of cut and machining allowance for traditional and optimized toolpath

points per line compared to the nodes of the FE model). Depths of cut and errors for this new discretization geometry are derived by linear interpolation of the values computed in the FE model nodes. By doing so, the computation cost is low (about 19 min in a laptop CPU 2.4 GHz Intel i5, RAM 4 GB for the whole optimization), while the resolution is high. Traditional and optimized toolpaths are presented in Fig. 6d, e.

The two toolpaths differ in both layers' selection and depths of cut. The proposed algorithm selects to machine each layer entirely before going on to the next and allows higher depths of cut on the final pass (around 1.5 mm against the 0.5 mm of the traditional toolpath). Moreover, as shown in Fig. 7 the selected depths of cut are not constant over the airfoil profile, indeed higher radial depths of cut are selected on the stiffer points (thicker part of the blade) and lower ones on the more flexible parts (trailing edge). Due to the evolution of the machining allowance (Fig. 7c), the proposed algorithm plans (multiple?) passes only on specific points (the more compliant), as shown in the example of Fig. 7a, while traditional approach consists in constant depth of cut passes, created by offsetting the airfoil geometry. In addition to this, as the tool goes down along the Z-axis, the number of passes at each layer decreases, since the stiffness of the component increases (Fig. 7c), proving that this strategy is in accordance with practical tips for thin-wall machining [16]. Thanks to the higher engagement conditions, the optimized toolpath machining cycle is faster than the traditional one: about 40 s against 90 s.

3.2 Machined Surface Results

To completely analyse the algorithm behaviour, machining surfaces have been acquired on each layer after being cut (interrupting the machining cycle before starting the next layer). For the sake of brevity, the machining errors on only one

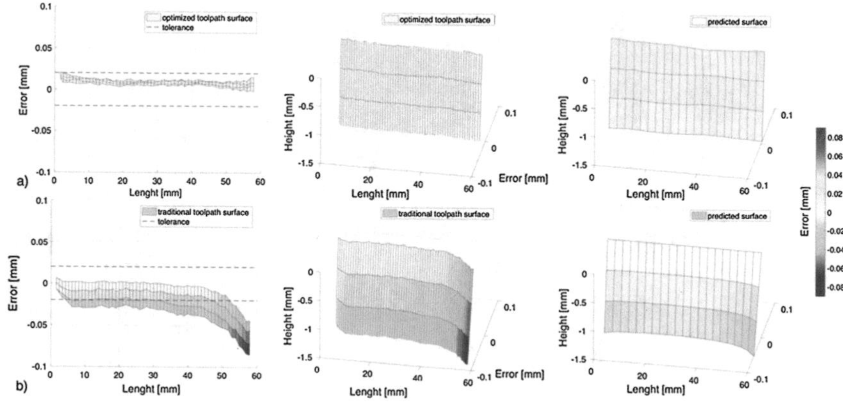

Fig. 8 Machining error on one side of the blade after machining the first layer **a** optimized toolpath, **b** traditional toolpath

side of the blade are presented (the results on the other side returned the same trend with opposite sign, i.e., the error on the thickness is double). Errors generated by the passes of the first layer (0/−5 mm) are presented in Fig. 8, for both traditional and optimized toolpath, including predicted values. For probe size reasons, only 1.5 mm could be measured. On the analyzed side of the blade, a negative error means a surface thicker than the nominal one, while positive means an overcutting of the surface.

As shown in Fig. 8a, optimized toolpath results in a low error surface (inside the tolerance range) almost constant all over the geometry, with good match between predicted and experimental errors. On the contrary, the first layer using the traditional toolpath (Fig. 8b) shows errors outside the tolerance range with higher value on the leading edge (i.e., the most flexible part). This trend was expected, since during down-milling, tool and workpiece static deflection causes an under-cut of the surfaces, not compensated by the traditional approach. Indeed, predicted surface is in line with the experimental results trend, even if an underestimation of the negative error on the leading edge is found. As soon as the most flexible layer is machined, the proposed approach manages to keep the error between the target tolerance (±0.02 mm), while a different toolpath, that does not consider the flexibility of the system, generates higher errors (almost 0.1 mm).

In Fig. 9, the same analysis is reported on measured surfaces after machining the second layer (−5/−10 mm), measuring 0/−6.5 mm range. Optimized toolpath (Fig. 9a) after two layers keeps the same trend of the previous case: results are inside the tolerance with a peak at around 4 mm from the top. Even so, predicted values differ from experimental ones: errors are higher, and the peak is around 5 mm from the top. This difference is found also on the traditional toolpath. Thus, the traditional toolpath errors are still higher and outside the tolerance range but lower than the one in the previous case, even in the zone machined in the first layer (0–5 mm). This is caused by the second layer passes over the previous machined surface. Indeed, since the previous machined surface was thicker that the one desired, the second pass will

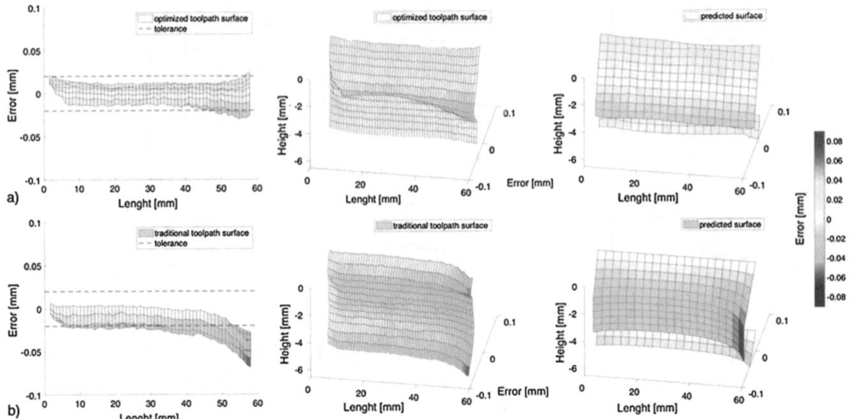

Fig. 9 Machining error on one side of the blade after machining the second layer **a** optimized toolpath, **b** traditional toolpath

also cut part of the previous surface. This effect is not considered by the algorithm, resulting in differences between predicted and actual surface errors. Nevertheless, after two layers the proposed algorithm can return a surface in tolerance, as desired, in contrast with the traditional toolpath that, even with this auto-leveling approach, fails to obtain a surface in tolerance. In fact, also the second layer presents a high compliance, resulting in a significant error on the surface.

This auto-leveling effect is even more advantageous in the machining of the whole case study, since the last passes on the stiffest part of the blade will affect the entire surface, reducing the actual error. This aspect is shown in Fig. 10, where the results

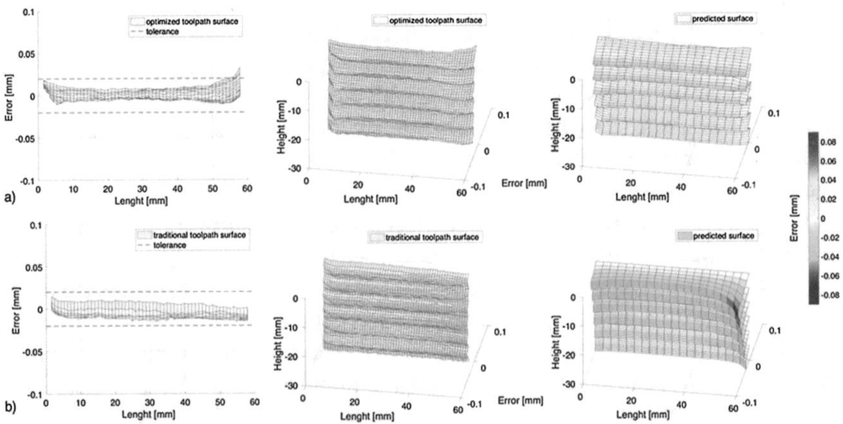

Fig. 10 Machining error on one side of the blade after the full machining cycle **a** optimized toolpath, **b** traditional toolpath

for the surface measured after the whole machining cycle (0/−40 mm) are presented (measures on 0/−30 mm range).

Figure 10 shows that, after the whole machining cycle is completed, both toolpaths produce surfaces inside the tolerance. Indeed, the machining errors of the traditional toolpath created by the passes on the first and the second layers have been drastically reduced by the passes on the following layers: the final passes on the last layer cut the oversized part of the blade in the upper zone, leveling the final surface. The effect is significant for multi passes 3-axis milling operations, especially in down-milling in which errors on the surface caused by deflection and cutting force lead to an under-cut of the surface (i.e., thicker part). On the contrary, single pass contouring operations and up-milling process should not be influenced by this effect. The proposed algorithm for error prediction (Sect. 2.3) does not include this effect, hence it overestimates the error for both toolpaths, however the method still provides an optimized toolpath that meets the target tolerance, reducing the number of passes and the cutting time.

4 Conclusions

In this paper, a technique to build the toolpath for 3-axis milling operations of thin-walled components is proposed. The method is the first step of the development of an integrated and automatic approach for toolpath generation, aiming to consider the flexibility of both the tool and the workpiece to select cutting parameters (i.e., engagement conditions), overcoming the limits of CAM software, that considers only the kinematic of the process. In this work, this procedure is developed for 3-axis milling: the algorithm generates the optimized toolpath, selecting the best cutting sequence (layers sequence and points path) to reduce the deflection of the component and computes the optimized radial depth of cut to keep the surface error in tolerance, given the other cutting parameters (e.g., axial depth of cut and feed). The method was experimentally validated and found to be accurate in predicting surface errors and computing the optimized toolpath in a single pass, reaching the target tolerance, while a traditional approach fails. On multi-passes, experimental results show that, in 3-axis down-milling, an auto-levelling effect was found. This effect is not included in the proposed approach and limits its application to single-pass contouring milling operations, where marks of the axial depth of cut are not acceptable, or multi-passes in up-milling process (where deflection causes an over-cut of the surface).

The presented technique lays the ground for the development of a more general approach, aiming at considering the flexibility of both the tool and the workpiece in the generation of the milling cycle. In addition to the radial depth of cut, the method could include the axial depth of cut and the feed to optimize the toolpath. Moreover, the method could be extended to simulate both more complex operations, such as 5-axis milling and additional effects, such as forced vibrations and unstable vibrations, since the FE model can be used to predict workpiece dynamics.

Acknowledgements The authors would like to thank Machine Tool Technology Research Foundation (MTTRF) and its supporters for the loaned machine tool (DMG MORI DMU 75 MonoBlock).

References

1. Sagherian R, Elbestawi MA (1990) A simulation system for improving machining accuracy in milling. Comput Ind 14(4):293–305
2. Budak E, Tunç LT, Alan S, Özgüven HN (2012) Prediction of workpiece dynamics and its effects on chatter stability in milling. CIRP Ann 61(1):339–342
3. Budak E, Altintas Y (1995) Modeling and avoidance of static form errors in peripheral milling of plates. Int J Mach Tools Manuf 35(3):459–476
4. Kolluru K, Axinte D (2013) Coupled interaction of dynamic responses of tool and workpiece in thin wall milling. J Mater Process Technol 213(9):1565–1574
5. Huang N, Yin C, Liang L, Hu J, Wu S (2018) Error compensation for machining of large thin-walled part with sculptured surface based on on-machine measurement. Int J Adv Manuf Technol 96(9):4345–4352
6. Rao VS, Rao PVM (2006) Tool deflection compensation in peripheral milling of curved geometries. Int J Mach Tools Manuf 46(15):2036–2043
7. Ratchev S, Liu S, Becker AA (2005) Error compensation strategy in milling flexible thin-wall parts. J Mater Process Technol 162–163:673–681
8. Wang J, Ibaraki S, Matsubara A (2017) A cutting sequence optimization algorithm to reduce the workpiece deformation in thin-wall machining. Precis Eng 50:506–514
9. Koike Y, Matsubara A, Yamaji I (2013) Design method of material removal process for minimizing workpiece displacement at cutting point. CIRP Ann 62(1):419–422
10. Desai KA, Rao PVM (2012) On cutter deflection surface errors in peripheral milling. J Mater Process Technol 212(11):2443–2454
11. Ratchev S, Liu S, Huang W, Becker AA (2006) An advanced FEA based force induced error compensation strategy in milling. Int J Mach Tools Manuf 46(5):542–551
12. Budak E (2006) Analytical models for high performance milling. Part I: cutting forces, structural deformations and tolerance integrity. Int J Mach Tools Manuf 46(12–13):1478–1488
13. Bolsunovskiy S, Vermel V, Gubanov G, Kacharava I, Kudryashov A (2013) Thin-walled part machining process parameters optimization based on finite-element modeling of workpiece vibrations. Procedia CIRP 8:276–280
14. Tuysuz O, Altintas Y (2017) Frequency domain updating of thin-walled workpiece dynamics using reduced order substructuring method in machining. J Manuf Sci Eng 139(7):71013–71016
15. Software Corporation MSC (2010) MD/MSC Nastran 2010 quick reference guide
16. Altintas Y (2012) Manufacturing automation: metal cutting mechanics, machine tool vibrations, and CNC design

Energy Efficient State Control of Machine Tool Components: A Multi-sleep Control Policy

Nicla Frigerio and Andrea Matta

Abstract Energy efficient control policies that switch off/on machine tools aim to reduce the energy consumed while not producing parts. Commonly, a transitory is required before resuming the service; thus, to switch off/on the machine might be not advantageous. This paper analyzes a time-based threshold policy that disables/enables machine components with separated control commands instead of controlling the whole machine simultaneously. According to the subset of controlled components, multiple sleeping states and transitory duration can be defined. Machine idle times are assumed stochastic and the expected value of the energy consumed per produced part is reduced while assuring a certain target of machine utilization. A simulation optimization algorithm is used to search for the optimum. The analysis is based on two real CNC machining centers. Potential benefits are compared to state-of-the-art policies and discussed for a set of realistic numerical cases representing several production environments.

Keywords Manufacturing automation · Energy efficient control · Multiple sleeping states

1 Introduction

Strategies that switch off/on production equipment in manufacturing environments have been recently proposed in the literature. These measures address the problem of energy efficiency at machine level focusing onto the reduction of the *non-processing energy (NPE)* [1]. This energy is usually denoted as *fixed energy* or *base load* and

N. Frigerio (✉) · A. Matta
Department of Mechanical Engineering, Politecnico di Milano, via G. la Masa 1, 20156 Milano, Italy
e-mail: nicla.frigerio@polimi.it

A. Matta
e-mail: andrea.matta@polimi.it

E. Ceretti and T. Tolio (eds.), *Selected Topics in Manufacturing*, Lecture Notes in Mechanical Engineering,
https://doi.org/10.1007/978-3-030-57729-2_4

it is related to the power requests of some machine components that keep executing their functions although the machine is not producing. For example, auxiliaries (e.g., chiller unit, hydraulic unit) allow to keep the machine in a *ready-for-process* state enabling machine tool cooling/heating, waste handling, and other machine condition such that whenever a part arrives, the part program can immediately start. For example, the chiller unit keeps the machine within a certain temperature range $[T_\ell, T_h]$ to assure process quality. A switching off command can trigger the machine in a low energy consumption state, i.e., a *sleeping* state, where some machine components are disabled. In order to resume the service, machine tools commonly need to visit a *startup* transitory state that begins with a switching on command. While in *startup*, a specific procedure is executed to reach the proper physical and thermal condition for working, such that the quality of processed parts is guaranteed. All machine components are active at procedure completion and the machine is *ready-for-process*. The described switching control does not affect the energy consumed while processing a part.

Control parameters to trigger the switching off/on commands must be properly selected to obtain a reduction of the NPE without compromising machine productivity. Although state control strategies are promising, in several cases the control is not advantageous because of significant startup energy or too long startup duration. Actually, during the *startup* transitory, it might happen that one component ends its own procedure earlier than another. For example, the hydraulic unit might establish the proper pressure in the oil circuit before the chiller unit has stabilized the temperature.

Figure 1 represents the behavior of some machine components when an energy efficient control is applied at machine level and components are switched off/on simultaneously. Startup procedure of component i has a duration y_i and machine startup ends when module $i = 2$ completes the startup procedure. Indeed, the service is resumed only when all components are *Enabled*, i.e., active and able to perform their functions upon request. Assuming that machine components can be controlled separately, a proper sequence of switching on commands can allow a synchronized startup completion improving machine energy efficiency. Also, it might happen that the best policy is to control only a subset of components according to machine

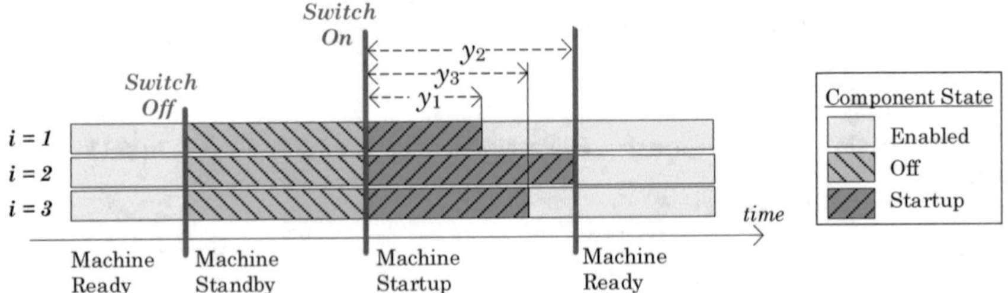

Fig. 1 States of components $i = \{1, 2, 3\}$ with switch off/on control applied at machine level. Components have different startup duration $y_i|\forall i$

characteristic and environmental condition (i.e., power requests, machine utilization, arrival process, productivity constraints). This paper considers a control policy that disables/enables machine components with separated control commands instead of controlling the whole machine. Following a bottom-up approach [2], machine model is obtained as combination of component models. As a consequence, the controlled machine might visit several *sleeping* states, each defined by a combination of component states.

1.1 Brief State of the Art on Energy Efficient Control

Relevant articles having the objective of reducing the NPE with an energy-efficient control (EEC) of machine state belong to the last control level in the Production Planning and Control hierarchy. EEC literature provides policies at machine level during production progress by assuming that arrivals are stochastic. The optimization problem usually incorporates an energy efficiency criterion in the objective function, a minimum productivity target as second objective or as a constraint, and it sometimes includes a reward for producing parts or serving customers.

A not-controlled machine is: *busy* when is working on parts, *breakdown* when a failure occurs and it needs to be repaired, and *idle* elsewhere. Sometimes, the machine is also considered as *blocked* when the downstream buffer is full. When an EEC is applied, the *idle* state is split into the following states: *sleeping* when the machine is partially deactivated, *ready-for-process* when all machine components are active, *startup* and *closedown* when the machine is executing a procedure to respectively activate and deactivate components. *Startup* and *closedown* are transitory states from/to *sleeping* and *ready-for-process* states. Despite models used in literature might have different assumptions, the *startup* cannot be neglected for most of manufacturing equipment and all works mentioned in this section consider the *startup* state, although only a subset considers also a *closedown* transitory [3–5].

A first group of studies [3, 6–8] analyses machine state control problems where the service is interrupted and resumed based on time information. Under a time-based policy, machines are controlled during starvation periods, i.e., waiting for parts. A second group analyses machine state control problems where the service is interrupted and resumed based on the number of accumulated parts in buffers such that machines can be controlled both during starvation periods [8, 9], blocking periods, or both [4, 10–13]. A combination of time and buffer information is used in a third group of studies [5, 14–16]. Although in literature the buffer-based policies are the most efficient as system complexity increases, time-based policies use a lower amount of information and can easily be applied to real production environments.

According to the literature, only one machine *sleeping* state is considered. As exceptions, multiple sleeping states are defined in Mashaei and Lennartson [16], Li and Sun [17] and Squeo et al. [18]. In Mashaei and Lennartson [16], machines are assumed to have two sleeping states: Hot Idle and Cold Idle. However, only the Cold Idle mode requires a transitory and the problem degenerates into a single-sleeping state problem. Li and Sun [17] assume that machines might have n sleeping states $(H_i|i = 1 \ldots n)$ where a closedown transitory is required to enter in state H_i as well as a startup to resume the service. Given a certain production scenario, the control problem chooses the most advantageous state H_i to use whenever the machine in a production line is idle or blocked according the estimated starvation/blocking periods. The switch-on is not optimized and the startup begins whenever the machine input buffer is not empty and the downstream buffer is not full. Squeo et al. [18] addressed the problem of controlling machine tool components using time information. Although the latter [18] is similar to the proposed approach, the effect of the control onto machine utilization is not included in the optimization which makes their problem unconstrained.

1.2 Contribution

This paper analyzes a time-based threshold policy that disables/enables machine components with separated control commands instead of controlling the whole machine simultaneously. Technological feasibility of the control policy is discussed focusing onto machine tools executing machining operations. The optimal control problem under a *Multi-Sleep* (MS) time-based control policy is formalized where decision variables represent the time instants to switch off/on machine components. Energy efficiency and productivity goals are included into the optimization problem. Considering a certain target on machine utilization, the optimal solution is obtained with a genetic algorithm and discrete event simulation is used to evaluate machine utilization and expected energy consumption. The numerical analysis is based on two real CNC machining centers. Potential benefits are compared to that of state-of-the-art policies and discussed for a set of scenarios representing several production environments.

2 Components Model and Control

2.1 Functional Mapping of Components

According to ISO14955-1:2017, the functional description of a machine tool is a general approach that shall identify relevant functions in terms of energy supplied to the machine. Focusing onto machines executing machining operations, functions are described in Table 1 and the assignment of functions to machine components follows. It might happen that a certain component is assigned to several functions. This generalized approach applies for a wide range of machine tools to evaluate their environmental impacts.

Table 1 Machine tool functions according to ISO14955-1:2017 and example of assigned components

Function	Description	Example of components assigned
Machine tool operation	Enable the primary machining process (cutting), motion and control	Spindles, linear and rotary axes (with their drives and power supply systems), PLCs, sensors, decoders/encoders
Process conditioning	Enable cooling, heating and other conditioning that are process-related in order to keep the process temperature and other relevant condition within limits	Cutting fluid pumps, cutting fluid cooler, lubrication fluid cooler
Workpiece and tool handling	Enable to storage, handle and clamp the workpiece and the tools	Tool magazine, compressed air system, workpiece/tool changing and clamping systems
Recyclables and waste handling	Enable to handle chips, scraps, and other waste materials	Chip conveyor, filter system, exhaust system
Machine tool cooling/heating	Enable cooling, heating and other conditioning that are independent of the machining process such that machine tool components are not damaged or distorted	Fans, air conditioning system, water cooler, cooling pumps, cooling/heating of guideways

2.2 *Energy States of Controllable Components*

Four energy states s are defined to characterize the behavior of components: *Off* ($s = 1$), *Hold* ($s = 2$), *Startup* ($s = 3$) and *On* ($s = 4$). The *Closedown* transitory state before entering in *Off* is not considered in this work because its duration is often negligible. In details:

- *Off*: the component is disabled and does not require any power;
- *Hold*: the component is enabled but is not executing any function;
- *On*: the component is enabled and is executing its function;
- *Startup*: the component is executing a procedure to pass from the *Off* state to the *Hold* state.

For example, the spindle in Fig. 2a becomes *On* when the part-program requires a certain cutting speed and it starts to accelerate, it returns into *Hold* state after deceleration. The *On* state can be further divided to reach a higher level of detail: *machining*, *acceleration*, or *air cutting* states. The spindle in Fig. 2a does not require any startup procedure to enter in *Hold* state and the transition is immediate.

Most of components presents a behavior similar to that of Fig. 2a and becomes *On* upon a specific request from machine control (PLC). Some components do not present a *Hold* state and pass from *Off* (or *Startup*) to *On*. Denote $c_{s,i}$ the power request of

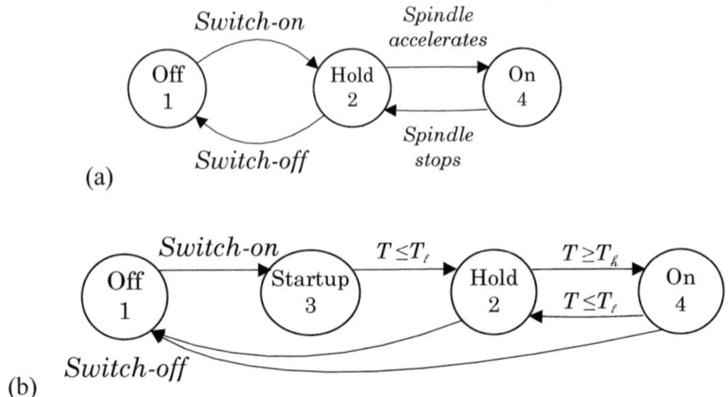

Fig. 2 Energy state models of **a** spindle and **b** chiller unit. The chiller has a periodic behavior

component i in state s. Trivially, the power consumed in *Off* is null ($c_{1,i} = 0$), the power in *On* state and in *Startup* is higher than that in *Hold* ($c_{4,i} > c_{2,i} \geq 0$ and $c_{3,i} > c_{2,i} \geq 0$).

Further, a last group of components presents a periodical behavior and visits *On* and *Hold* states according to an internal logic. As an example, the chiller (Fig. 2b) moves from *Hold* to *On* whenever machine temperature T raises above T_h, and moves from *On* to *Hold* when lower limit T_ℓ is reached. At switch on, the chiller directly enters in *Startup* state to assure target temperature before entering in *Hold* state. This transitory has a similar power request than that in *On* state ($c_{4,i} \approx c_{3,i}$) for all components belonging to this group. A periodic behavior is common for peripheral units and some machine auxiliaries whose operation depends on additional condition (e.g., machine and environment temperature, pressure).

The aggregation of *Hold* and *On* state represents the *Enabled* state of components (cf. Fig. 1) where all components are active and able to execute their function upon request.

2.3 Energy Model for a Controlled Machine

Let us consider a general machine composed by components $\mathbb{C}$ where a subset $\mathbb{I} \subseteq \mathbb{C}$ is controllable. Not controllable components stay Enabled (i.e. Hold and On). The machine is *busy* during part processing (including machine failures) and it becomes *idle* while waiting for new parts. Herewith, only machine idle states are modeled because the *busy* state is not affected by the control policy and because we assume that the machine cannot be blocked. Machine can be in the following *idle* states:

- *Ready-for-process* (or on-service): all machine components are enabled, and the process can start. Machine power request is $w_r = \sum \overline{c}_i + c_0$. The average power consumed while component $i \in \mathbb{I}$ is *Enabled* (*On* or Hold) is denoted as $\overline{c}_i$;

- *Sleeping*: a subset of machine components $i \in \mathbb{I}$ is *Off* or in *Startup* and the machine power request depends from such subset, thus a set of *sleeping states* is created;
- *Standby*: a particular sleeping state where all controllable components $i \in \mathbb{I}$ are *Off*. The machine power request is c_0 because of not controllable component power demand.

Assume that all components $\mathbb{C}$ are switched on/off simultaneously (cf. Fig. 1). The machine passes from *ready-for-process* to *standby* with the switch off. Components enter in *Startup* state simultaneously with the switch on but might finish at different points in time. At machine level, the startup transitory is actually a sequence of machine *sleeping* states with components in *Startup* or *Enabled* states. The duration of machine transitory is the longest startup duration of components. When all machine components are *Enabled*, the machine is *Ready-for-process*.

3 Control Problem for Energy Efficiency

A single machine working a single part type is considered such that the processing time p is assumed to be deterministic. The arrival of parts at machine is managed by an upstream mechanism such that a part is sent to the machine only during idling periods. Therefore, part interarrival times are $I = p + X$, where $X \in \mathbb{R}_0^+$ represents the time required by the mechanism to feed the machine. It is assumed that the mechanism may fail and therefore X is a random variable representing machine idle times and it is distributed according a probability density function $f(x)$ whose mean t_a is known. It is assumed that idle times are not affected by the control and this modelling choice yield to approximated results in different production cases.

It is assumed that the controlled machine cannot be blocked and is perfectly reliable. The latter assumption can be relaxed without requiring large extensions to the developed analysis. As discussed in Sect. 2, the machine includes a set $\mathbb{C}$ of components where subset $\mathbb{I} \subseteq \mathbb{C}$ is controllable. The subset $\overline{\mathbb{I} \cap \mathbb{C}}$ of not controllable components requires a fixed power c_0. Also, duration y_i for component startups is assumed deterministic. Although startup duration might vary according to machine condition and parameters imposed to assure part quality requirements, the range of variation is small enough to validate the latter assumption.

As common practice, the machine is always kept *ready-for-process* while waiting for new parts and it becomes *busy* during part processing. Denote this policy as **Always On (AON)** policy. Let us assume that components $i \in \mathbb{I}$ can be controlled with individual commands. A switch-off command triggers a component into the *Off* state and a switch on command might start component *Startup*. The component is *Enabled* (*Hold* or *On*) at startup end. The following policy is proposed such that time thresholds $\tau_{\text{off},i}$ and $\tau_{\text{on},i}$ are used to control component i:

Multi-Sleep policy (MS): Switch off the component when $\tau_{\text{off},i}$ has elapsed from the last departure. Switch on when $\tau_{\text{on},i} > \tau_{\text{off},i}$ has elapsed from the last departure or when next part arrives.

Denote the vector of control parameters $\boldsymbol{\tau} = \left\{\tau_{\text{off},i}, \tau_{\text{on},i}\right\}|i \in \mathbb{I}$. A component is never switched off when $\tau_{\text{off},i} = \infty$, and it is switched on upon part arrival when $\tau_{\text{on},i} = \infty$. The following optimization problem is formulated:

$$\min_{\boldsymbol{\tau}} Z(\boldsymbol{\tau}) = \phi(\boldsymbol{\tau}) + \alpha \cdot max\{0, \psi_{TARGET} - \psi(\boldsymbol{\tau})\} \tag{1}$$

$$subject\ to\text{:}\ \mathbf{A} \cdot \boldsymbol{\tau} < 0 \tag{2}$$

$$\tau_{off,i}, \tau_{on,i} \in \mathbb{R}_0^+;\ \ \forall i \in \mathbb{I} \tag{3}$$

Problem objective in Eq. (1) is the minimization of two terms. The first term $\phi(\boldsymbol{\tau})$ represents the expected energy (NPE) consumed by the machine per part produced. The second term represents a penalty term whenever the expected machine utilization $\psi(\boldsymbol{\tau})$ does not meet a certain minimum target ψ_{TARGET}. The weight $\alpha \in \mathbb{R}_0^+$ penalizes solutions where $\psi(\boldsymbol{\tau}) < \psi_{\text{TARGET}}$ such that as α increases, the second term becomes more important. On the contrary, when α decreases, the first term becomes more important and the energy is minimized while accepting machine utilization to be below the target. The extreme case where $\alpha = 0$ represents a problem without any utilization constraint. Constraints (3) define the domain of decision variables. Linear constraints (2) represent feasibility constraint existing among control parameters such that each switch-on command must happen after the associated switch-off command: $\tau_{\text{on},i} > \tau_{\text{off},i}|\forall i$. Matrix **A** also includes any switching precedence among components. For example, the spindle i cannot be activated before motor chiller k to prevent an excessive heat storage: $\tau_{\text{off},i} \leq \tau_{\text{off},k}$ and $\tau_{\text{on},i} \geq \tau_{\text{on},k}$. Also, the water pump and the compressor of a cooling unit must be activated simultaneously to allow a proper functioning of the thermal exchange.

Let us define a *cycle* as the time interval starting from the departure of a part and the departure of the next one. The cycle starts at $t = 0$ with all components *Enabled* and waiting for the part arrival. Denote x the arrival occurrence within a cycle, i.e., the realization of random variable X. Considering machine component i in isolation, four events $A_{i,k}|k = \{1, 2, 3, 4\}$ might happen. Whenever event $A_{i,k}$ happens, the energy $g_{i,k}$ consumed by component i is defined as in Table 2. Further, given the probability of occurrence $P(A_{i,k})$, the expected energy consumption in isolation $\varphi_i(\boldsymbol{\tau})$, of machine component i, is:$\varphi_i(\boldsymbol{\tau}) = \sum_k g_{i,k} \cdot P\left(A_{i,k}\right)$.

As in Fig. 1, it might happen that component i must wait for another component readiness consuming additional energy. Denote T_i as the time instant of startup completion for component i. T_i varies according to events $A_{i,k}$ because, for instance, the occurrence X can found the component "Enabled" if $x \leq \tau_{\text{off},i}$ or "Off" if $\tau_{\text{off},i} > x \geq \tau_{\text{on},i}$.

Table 2 Energy consumption of component i in isolation

Event $A_{i,k}$	Energy consumption of component i when event $A_{i,k}$ occurs
$A_{i,1} : x \leq \tau_{\text{off},i}$	$g_{i,1} = \overline{c}_i \cdot E[X|P(A_{i,1})]$
$A_{i,2} : \tau_{\text{off},i} < x \leq \tau_{\text{on},i}$	$g_{i,2} = \overline{c}_i \cdot \tau_{\text{off},i} + c_{3,i} \cdot y_i$
$A_{i,3} : \tau_{\text{on},i} < x \leq \tau_{\text{on},i} + y_i$	$g_{i,3} = \overline{c}_i \cdot \tau_{\text{off},i} + c_{3,i} \cdot y_i$
$A_{i,4} : x > \tau_{\text{on},i} + y_i$	$g_{i,4} = \overline{c}_i \cdot E[X|P(A_{i,4})] + \overline{c}_i(\tau_{\text{off},i} - \tau_{\text{on},i} - y_i) + c_{3,i} \cdot y_i$

Machine readiness is achieved at $T_j = \max T_i$, i.e., when all components has completed their own startup procedure. Therefore, T_j is random and component j is the last completing the startup, which might vary at each cycle.

Each component $i \neq j$ consumes an additional expected energy while waiting for machine readiness. Similarly, not controllable components keep absorbing power c_0 and consumes the additional energy $\varphi_0(\boldsymbol{\tau}) = c_0 \cdot E[T_j]$. Therefore, the expected machine energy consumption is:

$$\phi(\boldsymbol{\tau}) = \varphi_0(\boldsymbol{\tau}) + \sum_i [\varphi_i(\boldsymbol{\tau}) + \overline{c}_i \cdot E[T_j - T_i]] \tag{4}$$

4 Optimization Procedure

Matlab environment is used for implementation and the Genetic Algorithm (GA) embedded in Matlab is used to search for the optimum. Machine performance under a certain control policy is estimated using discrete event simulation: R independent replications are considered and each run includes N cycles (parts). These parameters are chosen to assure at most a half width of around 2% of the objective function obtained with AON policy. The tuning of GA parameters is described in Sect. 4.1. Common random number are used to evaluate candidate solutions among experiments. Let us consider a fictitious machine composed by four controllable and independent components with equal power requests: $\overline{c}_i = 2$ kW and $c_{3,i} = 2.4$ kW. Component startup varies such that $y_1 = 0$, $y_2 = 5$ s, $y_3 = 10$ s, and $y_4 = 30$ s. Machine processing time $p = 100$ s. Also, the comparison with Switching (SP) policy is considered. The SP policy has been proposed and analyzed in recent literature [6] such that component controls are simultaneous, i.e., $\tau_{\text{off},i} = \tau_{\text{off}}|\forall i$ and $\tau_{\text{on},i} = \tau_{on}|\forall i$.

4.1 Genetic Algorithm

A Weibull distribution with shape $k = 10$ and mean $t_a = 53.85$ s is considered for machine starvation times X such that machine utilization is 0.8 when the control is not applied. The computational budget is fixed at 4000 candidate evaluations. At each iteration, a population of candidates satisfying problem boundaries in constraints (2)–(3) is generated. GA selects candidates using a fitness scaling function based on candidate ranking. A certain elite is guaranteed to survive in next generation. New candidates are generated with a crossover function or with a mutation function and crossover ratio can be customized. Because of linear constraints in Eq. (2), the Arithmetic crossover function (children are the weighted arithmetic mean of two parents) and the Adaptive Feasible mutation function (the mutation chooses a direction and step length that satisfies bounds and linear constraints) are used. Some tests have been done to calibrate the parameters of the GA. Therefore, the following setting is selected for future experiments: the population size $PS = 50$, the elite fraction $EF = 0.05$, the crossover fraction $CF = 0.5$ and the fitness tolerance $FT = 10^{-6}$. The latter parameter is used to define whenever two candidate solutions perform equally.

4.2 Illustrative Example

A Weibull distribution with shape $k = 10$ and mean t_a is considered for machine starvation times X such that the stochastic process has Increasing Hazard Rate (IHR). Under this assumption, the arrival probability increases in time while approaching the mode of the distribution.

In order to represent cases with different machine utilization $u = p/(p + t_a)$, 8 scenarios are created by varying t_a. Each problem is solved using $R = 5$ and $N = 2500$ for simulation, and $\alpha = 0$ such that utilization target is relaxed. Results are collected in Table 3. The SP policy is advantageous compared to the AON policy for $u \in [0.6, 0.75]$ whilst for higher $u \in [0.8, 0.95]$ the SP degenerates into the AON policy, i.e., $\left\{\tau^*_{\text{off}}, \tau^*_{\text{on}}\right\} = \{\infty, \infty\}$. The MS policy, instead, is always the best policy. Indeed, MS choses the proper set of components to control keeping Enabled those with longer startup duration. In particular, component $i = 1$ is always controlled because it does not require startup, whilst component $i = 4$ is controlled only for high starvation times. This selective process allows MS to achieve a higher utilization than that obtained with SP policy.

Further, the MS policy results in a sequential startup. Optimal controls ($\tau^*_{\text{off},i}\, \tau^*_{\text{on},i}$ as in lines 9–10 of Table 3) show how the switch on commands are delayed according to the different startup duration y_i. The example of $u = 0.75$ is in Fig. 3: component $i = 4$ is kept Enabled, whilst components $i = 1, 2, 3$ are switched off at part departure ($\tau^*_{\text{off},i} = 0 | i = 1, 2, 3$) and switched on at different points in time. As

Table 3 Results of SP and MS policies for IHR (Weibull $k = 10$) starvation times varying t_a (CI95%)

	t_a (s)	66.67	53.85	42.86	33.33	25.00	17.65	11.11	5.26
AON	Z^{AO} (kJ/part)	533.3 ± 1.3	430.8 ± 0.9	342.7 ± 1.4	266.5 ± 0.4	200 ± 0.3	141.1 ± 0.4	88.9 ± 0.3	42.1 ± 0.2
	u	0.6	0.65	0.7	0.75	0.8	0.85	0.9	0.95
SP	$Z^{SP}(\tau^*)$ (kJ/part)	258 ± 0.1	258 ± 0.1	258 ± 0.1	258 ± 0.1	(As AO)	(As AO)	(As AO)	(As AO)
	$\psi(\tau^*)$	0.509	0.544	0.579	0.623	(As AO)	(As AO)	(As AO)	(As AO)
	$\tau^*_{off}; \tau^*_{on}$	0; ∞	0; ∞	0; ∞	0; ∞	∞; ∞	∞; ∞	∞; ∞	∞; ∞
MS	$Z^{MS}(\tau^*)$ (kJ/part)	135 ± 0.6	128.5 ± 0.4	123.8 ± 0.2	115.9 ± 0.3	96.3 ± 0.2	84.78 ± 0.1	60.5 ± 0.1	31.6 ± 0.1
	$\psi(\boldsymbol{\tau}^*)$	0.585	0.632	0.688	0.74	0.789	0.854	0.896	0.95
	$\tau^*_{off,1}; \tau^*_{on,1}$	0; ∞	0; ∞	0; ∞	0; ∞	0; ∞	0; ∞	0; ∞	0; ∞
	$\tau^*_{off,2}; \tau^*_{on,2}$	0; 64	0; 52	0; 41	0; 29	0; 21	0; 13	0; 6	∞; ∞
MS	$\tau^*_{off,3}; \tau^*_{on,3}$	0; 59	0; 47	0; 34	0; 24	0; 16	0; 8	∞; ∞	∞; ∞
	$\tau^*_{off,4}; \tau^*_{on,4}$	0; 39	0; 27	0; 14	∞; ∞	∞; ∞	∞; ∞	∞; ∞	∞; ∞

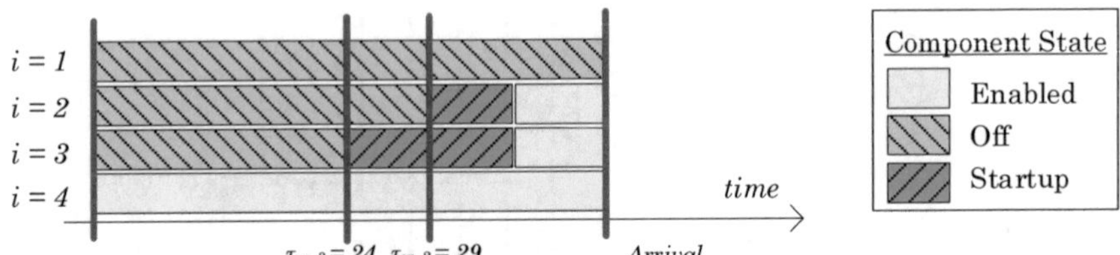

Fig. 3 States of machine components with MS policy (scenario with $u = 0.75$). At optimality, most of the arrivals occurs after the end of startup procedure executed by components $i = 2, 3$

a consequence, components 2 and 3 finish the startup simultaneously. Component $i = 1$ has no startup and it is switched on upon part arrival $\left(\tau^*_{\text{on},1} = \infty\right)$.

5 Real Case Application

In this section, two real machining centers are modelled and the feasibility of the control is assessed. For both machines, spindle and drives (spindle, axis, tool magazine) do not require power in machine idle states and they are not included in the model. Two versions of the problem are solved: in the first version, the weight $\alpha = 0$ such that the second term in Eq. (2) is neglected (unconstrained problem), in the second version, a high weight α is chosen ($\alpha = 10{,}000$) such that machine utilization target is met. In the following experiments, discrete event simulation is used with $N = 10{,}000$ and $R = 5$. The solver performs 4,000 iterations in order to search for solution and the computational time required to perform one iteration is on the average 0.06 s for SP policy and 0.08 s for MS policy. Results have been obtained on a laptop with i5 Intel Core @2.4 GHz and RAM-8 GB.

5.1 Machine Modeling and Experiment Description

The first machine (M1) is a 5-axis vertical machining center with 30 kW of installed power and 600 × 450 × 450 working cube. Motor chiller unit (spindle and axes), hydraulic unit and coolant extraction pump have a periodical behavior depending respectively on temperature, pressure, and coolant levels in machine basin. The waste handling components (i.e., two chip conveyors), and two coolant pumps (furnishing filtered cutting fluid at low pressure) are constantly On. Nominal power requests for components in On state ($c_{4,i}$) are in Table 4. Further, it is assumed that other powers are: $c_{2,i} = 0$, $c_{3,i} = c_{4,i}$, and $\bar{c}_i = 0.2 \cdot c_{4,i} | \forall i$. It is also assumed that not controllable components power is $c_0 = 0.25 \cdot \sum \bar{c}_i$. According to operator experience, it is assumed that a significant transitory belongs to chiller unit, hydraulic unit, and extraction pump. For each component $\tau_{\text{off},i} < \tau_{\text{on},i}$ and no technological precedence

Table 4 Machine 1 and Machine 2 parameters

Machine	Component i (or group of)	Startup	On	Hold	Enabled	y_i (s)
		$c_{4,i}$ (kW)	$c_{3,i}$ (kW)	$c_{2,i}$ (kW)	$\bar{c}_i$ (kW)	
M1	Motor chiller unit i = 1	2	2	0	0.6	30
M1	Hydraulic unit i = 2	0.75	0.75	0	0.225	5
M1	Coolant extraction pump i = 3	0.24	0.24	–	0.072	10
M1	Chip conveyors i = 4	–	0.24	–	0.24	0
M1	Coolant pumps i = 5	–	1.84	–	1.84	0
M2	Motor chiller unit i = 1	3.3	3.3	0.6	1.07	90
M2	Hydraulic unit i = 2	4.2	4.2	0.5	0.99	5
M2	Waste handling, coolant pumps, lights, displays i = 3	–	1.11	–	1.11	0
M2	Others i = 4	1.66	1.66	–	1.66	20

is required. Chip conveyor and coolant pumps can be controlled together with no impact on the results, since they have negligible startup. The device providing parts to the machine has a stochastic behavior and, thus, machine starvation times X follow a Weibull with shape $k = 0.7$ (i.e., Decreasing Hazard Rate): the arrival probability decreases in time indicating a possible failure of the feeder. Given $p = 100$ s, 8 scenarios with different machine utilizations are created by varying t_a. The second machine (M2) is a 5-axis horizontal machining center for powertrain applications with 110 kW of installed power and 700 × 700 × 800 working cube. Not controllable components are the numerical control system and the AC of the control cabinet. The waste handling components (e.g., exhaust air, chip conveyor), the coolant pumps, and other components (e.g., lights, displays) are constantly On. The hydraulic unit, and the motor chiller unit have a periodical behavior and their power footprints have been acquired with dedicated experimental measures (Table 4). Also, power data have been acquired at machine level: the subset of not controllable components absorbs $c_0 = 0.52$ kW and the machine absorbs on the average 5.35 kW while ready-for-process where 39% is associated to hydraulic and chiller units. Further, it is assumed that 21% (1.11 kW) is related to waste handling components, process coolant pumps, lights and display. These components have negligible startup and can be controlled together with no impact on the results. The remaining 31% (1.66 kW) is associated to the coolant filter and security systems that requires around 20 s to perform a startup procedure. For each component $\tau_{\text{off},i} < \tau_{\text{on},i}$ and no technological precedence is required. It is assumed that M2 is connected upstream with another machine, that it exists an intermediate buffer, and that the upstream machine can fail. Processing time is $p = 100$ s and occurrences at M2 happen after 2 s (deterministic loading time from the buffer) or according to a Weibull distribution with $k = 15$ and $t_a = 55$ s (due to upstream failures).

5.2 Results and Policy Comparison

The MS policy is compared with the AON and the SP policies in terms of energy and machine utilization. For machine M1, results are represented in Fig. 4. The SP policy enables to save energy when mean starvation time t_a is high, and converges to the AON policy as t_a decreases because machine utilization increases. The MS performs better in all analyzed cases. Chip conveyors and coolant pumps are always switched off/on because they do not require any transitory and therefore, MS achieves always a saving without compromising machine utilization.

When a target is included in the problem, the savings obtained with SP are limited to 1–2% of the energy, whereas MS results are not modified because the obtained utilization is already into the feasibility space or very close. As example, the MS solution obtained for $u_{AON} = 0.6$ ($t_a = 66.6$ s) is reported in Table 5. In the unconstrained case, the switch off of the hydraulic unit ($i = 2$) and the coolant extractor pump ($i = 3$) are delayed ($\tau^*_{\text{off},i} > 0 | i = 2, 3$) and the chiller unit is never switched off ($\tau^*_{\text{off},1} = \infty$). Chip conveyor and coolant pumps do not require startup and they are switched off at departure ($\tau^*_{\text{off},i} = 0 | i = 4, 5$). To assure the utilization target, the control of $i = 2$ changes from $\tau^*_{\text{off},2} = 34.5$ s to $\tau^*_{\text{off},2} = 70.7$ s. Since arrival probability decreases in time, to delay the switch off increases the

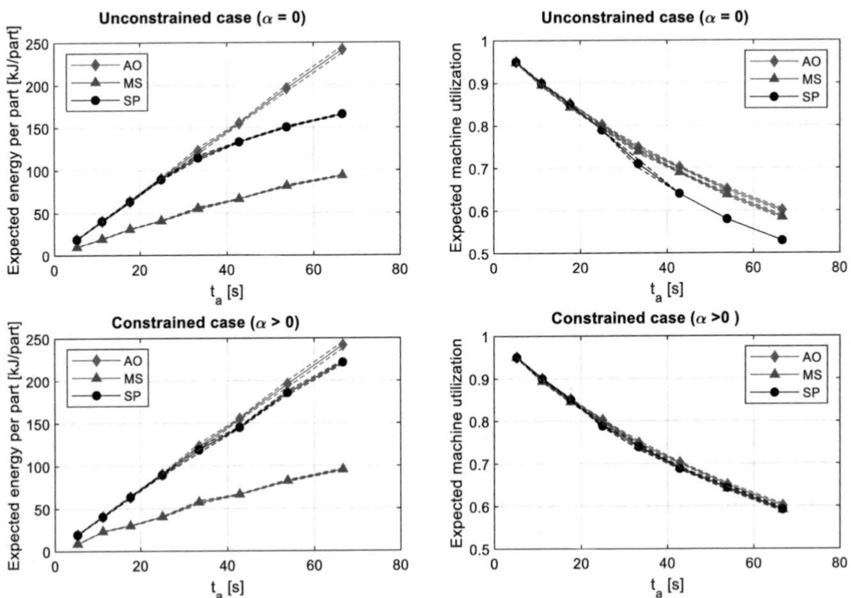

Fig. 4 Experiments on M1 for unconstrained (top panels) and constrained (bottom panels) optimizations. Expected energy and expected machine utilization obtained with AON, SP and MS policies are compared as the mean starvation time t_a increases. Dotted lines represent CI95%. In bottom panels, the target ψ_{TARGET} is the 98% of machine utilization obtained with the AON policy

Table 5 Results for Machine 1 in case with $t_a = 66.67$ s

Case	ψ_{TARGET}	$Z^{MS}(\tau^*)$ (kJ/part)	u	$\tau^*_{off,1}$; $\tau^*_{on,1}$	$\tau^*_{off,2}$; $\tau^*_{on,2}$	$\tau^*_{off,3}$; $\tau^*_{on,3}$	$\tau^*_{off,4}$; $\tau^*_{on,4}$	$\tau^*_{off,5}$; $\tau^*_{on,5}$
Unconstrained	0	93.76 ± 1.01	0.585 ± 0.004	∞; ∞	34.9 s; ∞	34.9 s; ∞	0; ∞	0; ∞
Constrained	0.588	95.52 ± 1.2	0.591 ± 0.004	∞; ∞	70.7 s; ∞	74.5 s; ∞	0; ∞	0; ∞

Table 6 Results for Machine 2

$Z^{MS}(\tau^*)$ (kJ/part)	u^{MS}	$\tau^*_{off,1}$; $\tau^*_{on,1}$	$\tau^*_{off,2}$; $\tau^*_{on,2}$	$\tau^*_{off,3}$; $\tau^*_{on,3}$	$\tau^*_{off,4}$; $\tau^*_{on,4}$
73.4 ± 1.6	0.793 ± 0.004	∞; ∞	2 s; ∞	0 s; ∞	2 s; 40.6 s

expected utilization. Similarly for $i = 3$. Components are switched on upon part arrival ($\tau^*_{on,i} = \infty | i = 2, 3, 4, 5$).

When machine M2 is not controlled, it consumes 124.7 ± 2.5 kJ/part with utilization $u = 0.811 \pm 0.003$. Because of the long startup duration of the chiller, the SP policy is not advantageous. The MS policy achieves 26.3% of saving by controlling all components except the chiller. Trivially for components not requiring startup ($i = 3$), solution $\{\tau^*_{off,3}; \tau^*_{on,3}\} = \{0; \infty\}$ is obtained. The hydraulic unit ($i = 2$) and components $i = 4$ are switched off with $\tau^*_{off,2} = \tau^*_{off,4} = 2$ s, when the arrival peak has elapsed. Because of the startup procedure, components $i = 4$ are switched on in advance such that the machine is ready when the probability of arrival increases again ($\tau^*_{on,4} = 40.6$ s). The hydraulic unit is switched on upon part arrival because its startup is short. This optimal MS control obtains 73.4 kJ/part (41% of saving) and also satisfies the target utilization of 0.79. Results are reported in Table 6.

6 Conclusive Remarks and Future Developments

Individual controls at component level can be applied for machine tools and numerical results show that considering multiple sleeping states can improve the performance of switching policies thanks to the selection of which component to switch. At optimality, several sleeping states are visited in sequence. Also, the selection of component subset avoids heavy startup procedures increasing the potential benefit of EEC policies under strict productivity targets. MS policy is advantageous when there is no benefit using the SP policy due to one or more components with significant startup energy or duration. Future effort will be devoted to develop more efficient optimization algorithm. A critical barrier for implementation remains the knowledge of the waiting time distribution that is assumed as known. Learning methods should be included to estimate machine starvation times and to increase applicability of the proposed policy for practitioners.

References

1. Dahmus JB, Gutowski TG (2010) An environmental analysis of machining. In: Manufacturing engineering and materials handling of engineers, pp 643–652
2. Frigerio N, Matta A, Ferrero L, Rusinà F (2013) Modeling energy states in machine tools: an automata based approach. In: Proceedings of the 20th CIRP international conference on life cycle engineering 2013, Singapore, pp 203–208

3. Mouzon G, Yildirim MB, Twomey J (2007) Operational methods for minimization of energy consumption of manufacturing equipment. Int J Prod Res (IJPR) 45(18–19):4247–4271
4. Jia Z, Zhang L, Arinez J, Xiao G (2016) Performance analysis for serial production lines with Bernoulli machines and real-time WIP based machine switch-on/off control. IJPR 54(21):6285–6301
5. Guo X, Zhou S, Niu Z, Kumar P (2013) Optimal wake-up mechanism for single base station with sleep mode. In: IEEE international teletraffic congress 2013, pp 1–8
6. Frigerio N, Matta A (2015) Energy efficient control strategies for machine tools with stochastic arrivals. IEEE Trans Autom Sci Eng (TASE) 12(1):50–61
7. Frigerio N, Matta A (2014) Energy efficient control strategy for machine tools with stochastic arrivals and time dependent warm-up. Procedia CIRP 15:56–61
8. Li W, Zein A, Kara S, Herrmann C (2011) An investigation into fixed energy consumption of machine tools. In: Proceedings of the 18th CIRP international conference on life cycle engineering, Braunschweig, pp 268–273
9. Frigerio N, Matta A (2016) Analysis on energy efficient switching of machine tool with stochastic arrivals and buffer information. IEEE Trans Autom Sci Eng 13(1):238–246
10. Su W, Xie X, Li J, Zheng L, Feng L (2017) Reducing energy consumption in serial production lines with Bernoulli reliability machines. IJPR 55(24):7356–7379
11. Wang J, Feng Y, Fei Z, Li S, Chang Q (2017) Markov chain based idle status control of stochastic machines for energy saving operation. In: IEEE international conference on automation science and engineering (CASE), Xian, pp 1019–1023
12. Li Y, Chang Q, Ni J, Brundage M (2018) Event-based supervisory control for energy efficient manufacturing systems. IEEE Trans Autom Sci Eng 15(1):92–103
13. Brundage MP, Chang Q, Li L, Xiao GX, Arinez J (2014) Energy efficiency management of an integrated serial production line and HVAC system. IEEE Trans Autom Sci Eng 11(3):789–797
14. Frigerio N, Matta A (2018) Analysis of production lines with switch off/on controlled machines. In: Thiede S, Herrmann C (eds) Eco-factories of the future. Springer International Publishing
15. Maccio VJ, Down DG (2015) On optimal policies for energy aware servers. Perform Eval 90:36–52
16. Mashaei M, Lennartson B (2013) Energy reduction in a pallet-constrained flow shop through on-off control of idle machines. IEEE Trans Autom Sci Eng 10(1):45–56
17. Li L, Sun Z (2013) Dynamic energy control for energy efficiency improvement of sustainable manufacturing systems using Markov decision process. IEEE Trans Syst Man Cybern 43(5):1195–1205
18. Squeo M, Frigerio N, Matta A (2019) Multiple sleeping states for energy saving in CNC machining centers. Procedia CIRP 80:144–149

μEDM Machining of ZrB_2-Based Ceramics Reinforced with SiC Fibres or Whiskers

Mariangela Quarto, Giuliano Bissacco, and Gianluca D'Urso

Abstract The effects of different reinforcement shapes on stability and repeatability of micro electrical discharge machining were experimentally investigated for Ultra-High Temperature Ceramics based on zirconium diboride (ZrB_2) doped by SiC. Two reinforcement shapes, namely SiC short fibres and SiC whiskers were selected in accordance with their potential effects on mechanical properties and oxidation performances. Specific sets of process parameters were defined minimizing the short circuits in order to identify the best combination for different pulse types. The obtained results were then correlated with the energy per single discharge and the discharges occurred for all the combinations of material and pulse type. The pulse characterization was performed by recording pulses data by means of an oscilloscope, while the surface characteristics were defined by a 3D reconstruction. The results indicated how reinforcement shapes affect the energy efficiency of the process and change the surface aspect.

Keywords micro-EDM · UHTCs · Ultra High Temperature Ceramics · Machinability

M. Quarto (✉) · G. D'Urso
Department of Management, Information and Production Engineering, University of Bergamo, Via Pasubio 7/b, 24044 Dalmine, BG, Italy
e-mail: mariangela.quarto@unibg.it

G. D'Urso
e-mail: gianluca.d-urso@unibg.it

G. Bissacco
Department of Mechanical Engineering, Technical University of Denmark, Produktionstorvet, Building 427, 2800 Kgs., Lyngby, Denmark
e-mail: gibi@mek.dtu.dk

E. Ceretti and T. Tolio (eds.), *Selected Topics in Manufacturing*, Lecture Notes in Mechanical Engineering,
https://doi.org/10.1007/978-3-030-57729-2_5

1 Introduction

Among the advanced ceramic materials, Ultra-High-Temperature ceramics (UHTCs) are characterized by excellent performances in extreme environment. This family of materials is based on borides (ZrB_2, HfB_2), carbide (ZrC, HfC, TaC) and nitrides (HfN), which are characterized by high melting point, high hardness and good resistance to oxidation in several environments. In particular, ZrB_2-based materials are of particular interest because of their suitable properties combination and are considered promising for several applications; for example, among the most attractive applications, one is in the aerospace sector as a component for the re-entry vehicles and devices [1–4].

The relative density of the base material ZrB_2 is usually about 85% because of the high level of porosity of the structure; furthermore, in the last years, researchers are focused on fabricating high-density composites characterized by good strength (500–1000 MPa). For these reasons, the use of single-phase materials is not sufficient for high-temperature structural applications. Many efforts have been done on ZrB_2-based composites in order to improve the mechanical properties, oxidation performances, and fracture toughness; however, the low fracture toughness remains one of the greatest efforts for the application of these materials under severe conditions [5–10]. Usually, the fracture toughness of ceramic materials can be improved by incorporating appropriate reinforcements that activate toughening mechanisms such as phase transformation, crack pinning, and deflection. An example is the addition of SiC; in fact, it has been widely proved that its addition improves the fracture strength and the oxidation resistance of ZrB_2-based materials due to the grain refinement and the formation of a protective silica-based protective layer. Based on these aspects, the behaviour of ZrB_2-based composites, generated by the addition of SiC with different shapes (e.g. whiskers or fibres), have been studied in recent works. It has been reported that the addition of whiskers or fibres gives promising results, improving the fracture toughness and this improvement could be justified by crack deflection [5, 6, 9, 11, 12]. The critical aspect of the reinforced process was the reaction or the degeneration of the reinforcement during the sintering process itself [13, 14].

Despite all the studies that aim to improve mechanical properties and resistance, this group of materials is very difficult to machine by means of traditional technologies, because of their high hardness and fragility. Only two groups of processes are effective in processing them: on one side the abrasive processes as grinding, ultrasonic machining, and waterjet, on the other side the thermal processes such as laser and electrical discharges machining (EDM) [15–18].

In this work, ZrB_2 materials containing 20% vol. SiC whiskers or fibres produced by hot pressing were machined by the μEDM process; in particular, the effect of the non-reactive reinforcement shapes on the process performances were investigated, verifying if the process is stable and repeatable for advanced ceramics. The choice of 20% vol. is related to the evidence reported in some previous works [14, 18, 19], in which it was shown that this fraction of reinforcement allowed generating the best combination of oxidation resistance and mechanical characteristics useful for obtaining better results in terms of process performances and dimensional accuracy for features machined by micro-EDM.

2 Materials

The following ZrB_2-based composites, provided by ISTEC-CNR of Faenza (Italy), were selected for evaluating the influence of reinforcement shape on the process performance of micro-slots machined by mEDM technology:

- ZrB_2 + 20% SiC short fibres, labelled as ZrB20f
- ZrB_2 + 20% SiC whiskers, labelled ZrB20w.

As reported by Silvestroni et al. [14], ZrB_2 Grade B (H.C. Starck, Goslar, Germany), SiC HI Nicalon-chopped short fibres, Si:C:O = 62:37:0.5, characterized by 15 μm diameter and 300 μm length or SiC whiskers characterized by average diameter 1 μm and average length 30 μm were used for the ceramic composites preparation.

The powder mixtures were ball milled for 24 h in pure ethanol using silicon carbide media. Subsequently, the slurries were dried in a rotary evaporator. Hot-pressing cycles were conducted in low vacuum (100 Pa) using an induction-heated graphite die with a uniaxial pressure of 30 MPa during the heating and were increased up to 50 MPa at 1700 °C (T_{MAX}), for the material containing fibres, and at 1650 °C (T_{MAX}) for the composites with whiskers. The maximum sintering temperature was set based on the shrinkage curve. Free cooling followed and details about the sintering runs are reported in Table 1, where T_{ON} identifies the temperature at which the shrinkage started. Density was estimated by the Archimedes method.

After the preparation, the raw materials were analysed by the Scanning Electron Microscope (SEM) (Fig. 1). The samples reinforced by the fibres show a very clear separation between the base matrix and the non-reactive reinforcement. Fibres dispersion into the matrix is homogenous and the fibres are characterized by the same orientation, as no agglomeration was observed in the sintered body. For the sample containing whiskers reinforcement, a dense microstructure was observed and the whiskers are generally well dispersed into the matrix. The SEM-EDS analysis

Table 1 Composition, sintering parameters of the hot-pressed samples [14]

Label	Composition	T_{ON} (°C)	T_{MAX} (°C)	Final density (g/cm^3)	Relative density (%)
ZrB20f	ZrB_2 + 20% SiC Short fibres	1545	1650	4.89	94.0
ZrB20w	ZrB_2 + 20% SiC Whiskers	1560	1700	5.22	97.0

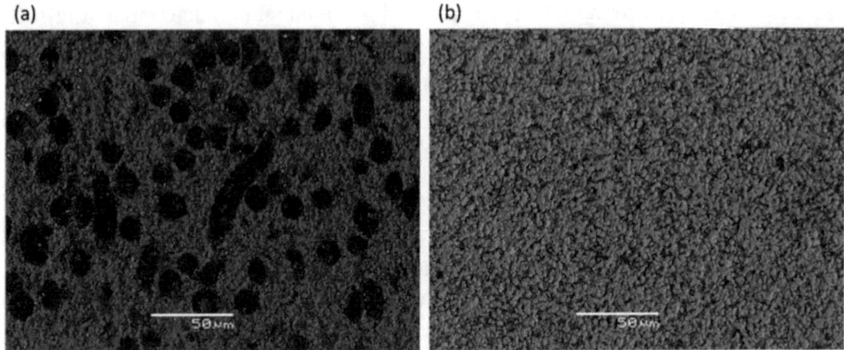

Fig. 1 SEM backscattered images of typical appearance of ZrB20f (**a**) and ZrB20w (**b**)

confirms these statements: for short fibres it is possible to observe a clear separation between the base matrix and the reinforcement. On the contrary, the analysis performed for samples containing whiskers shows that the two materials are mixed.

3 Electrical Discharge Machining Set-Up

3.1 Experimental Set-Up

The effects of selected reinforcement shapes were investigated by conducting μEDM milling experiments. A simple circular pocket having a diameter equal to 1 mm and a depth of about 200 μm was selected as the test feature. Solid tungsten carbide electrode with a diameter equal to 300 μm was used as a tool, and hydrocarbon oil was used as a dielectric fluid. The experiments were performed for three different process parameter settings, corresponding to different pulse shapes. The instantaneous values cannot be set for some process parameters, because the machine used for experiments presents an autoregulating system. Thus, the characterization of electrical discharges is very important, not only to assess the real value of process parameters but, most of all, to evaluate the stability and repeatability of the process. An acquisition system was developed and used to collect waveforms for characterizing the process. A current monitor with a bandwidth of 200 MHz and a voltage probe were connected to the μEDM machine and to a programmable counter and a digital oscilloscope (Rohde & Schwarz RTO1014). These connections allow to acquire the current waveforms and count the occurred discharges. The counter has been set once the trigger value was established to avoid recording and counting the background noise.

Preliminary tests were performed to define the optimal process parameters for each combination of material and pulse type. The experimental campaign was based on a general full factorial design, featured by two factors: the reinforcement shape, defined by two levels, and the pulse type, defined by three levels. Different levels of pulse types identify the different duration of the discharges; in particular, level A is referred to long pulses while level C identify the short pulses. Three repetitions were performed for each run.

3.2 Discharge Population Characterization

Discharge populations were characterized by repeated waveform samples of current and voltage signals. The current and voltage probes were connected to the digital oscilloscope having a real-time sample rate of 40 MSa/s. The trigger level of the current signal was set to 0.5 A in order to acquire all the effective discharges avoiding the background noise. The acquired waveform samples were stored in the oscilloscope buffer and then transferred to a computer to be processed by a Matlab code, written by the authors and shown schematically in Fig. 2.

The Matlab code was used to calculate the numbers of electrical discharges, their instantaneous values, duration (width) and voltage. Finally, the average value of energy per discharge (E) was estimated by integrating the instantaneous value of the power, calculated as the product of the instantaneous values of current (i(t)) and voltage (v(t)), with respect to the time (t). In Fig. 3 the frequency distribution histograms show the discharge population distributions as a function of reinforcement shape and pulse type applied to the machining. Histograms show a good reproducibility and stability of the process, providing information regarding the frequency waveforms with different peaks of current. The normal distribution well represents the discharge samples which suggesting a stable process. Considering both reinforcement shapes, the intensities of the pulses are included in a similar range for pulses type A and B, while, for short pulse, they are characterized by some differences: for whiskers, the maximum peak current corresponds to the average value of the peaks occurred for the short fibres.

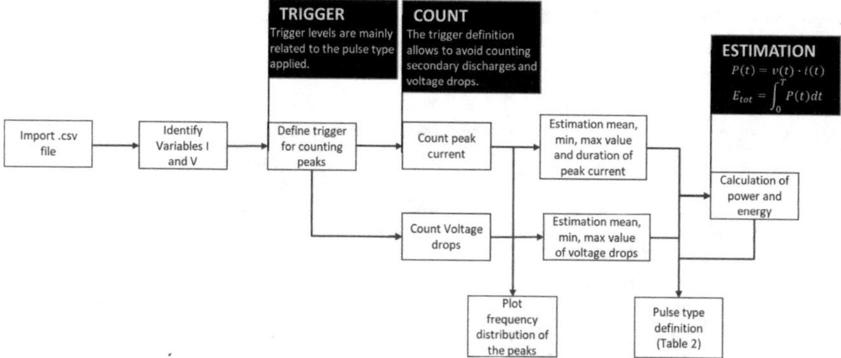

Fig. 2 Flowchart of Matlab code written by authors for the pulse characterization

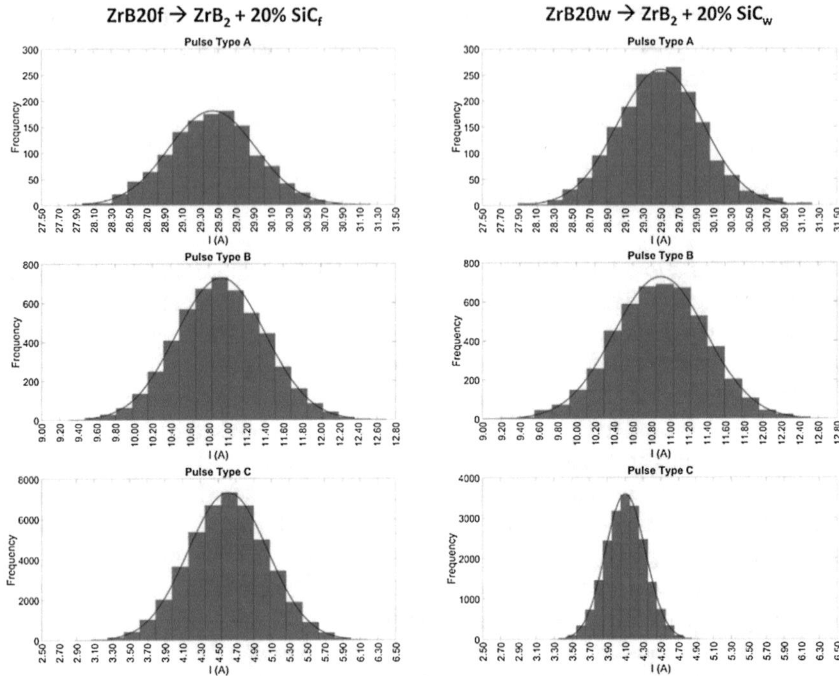

Fig. 3 Examples of frequency distribution histograms for pulses occurred during $ZrB_2 + 20\%SiC$ machining

$$E = \int_0^T v(t) \cdot i(t)dt \tag{1}$$

Through the Matlab data elaboration, it was possible to define the instantaneous value of the process parameters. The pulse characteristics for the estimated parameters are reported in Table 2.

Table 2 Mean values of the main parameters describing the pulse type

Reinforcement shape	Pulse type	Peak current (A)	Open circuit voltage (V)	Width (μs)	Energy per discharge (μJ)
Fibres (ZrB20f)	A	29.44	69.57	0.70	779.36
	B	10.93	92.89	0.33	150.30
	C	4.61	100.79	0.06	14.73
Whiskers (ZrB20w)	A	29.49	70.75	0.70	784.99
	B	10.91	91.70	0.33	161.00
	C	4.10	99.60	0.06	21.41

In terms of peaks of current and voltage, the differences between the two materials, machined by the same pulse type, are really tiny. The only relevant difference can be remarked for energy per discharge, in fact the discharges developed during the machining on the ZrB20w generate higher energy in comparison to the energy per discharge of ZrB20f. In particular, the energy per discharge generated by pulse type A for ZrB20w are 0.72% higher than the energy developed by the same pulse type for ZrB20f. This is a very tiny difference, but considering the pulse type C, the energy per discharge developed is about 45% higher than the energy generated by the same pulse type for fibres.

3.3 Characterization Procedure

A 3D reconstruction of micro-slots was performed by means of a confocal laser scanning microscope (Olympus LEXT) with a magnification of 20×. Then, the images were analysed with a scanning probe image processor software (SPIP by Image Metrology). A plane correction was performed on all the images to level the surfaces and to remove primary profiles; then, the surface roughness (Sa) was assessed by the real-topography method, based on the international standard UNI EN ISO 25178:2017. The process performance was evaluated through the estimation of two indicators related to the performance in terms of machining velocity and tool wear compared to the number of occurred discharges. The third indicator taken into account evaluates the tool wear ratio.

Material Removal per Discharge (MRD) was calculated as the ratio between the volume of micro-slots (MRW) estimated by the scanning probe image processor software and the number of discharges recorded by the programmable counter.

Since this kind of materials is characterized by high level of porosity, to get the actual values of MRD, the volume of the micro-slots was adjusted considering the relative density (δ) defined in Table 1. Taking into account the density allows compensating for the presence of porosity in the sample structure. Thus, the new indicator considered for the analysis was estimated as reported in (2).

$$MRD_{\delta} = MRD \cdot \delta \tag{2}$$

Tool Wear per Discharge (TWD) was estimated as the ratio between the volume of electrode wear (MRT) and the number of discharges recorded by the programmable counter. Tool wear was measured as the difference between the length of the electrode before and after the single milling machining. The length was measured through a touching procedure executed in a reference position. The electrode wear volume was estimated starting from the length of the tool wear and considering the tool as a cylindrical part. The Tool Wear Ratio (TWR) was calculated as the ratio between the previous performance indicators, considering the relative density of the workpiece material (TWR_{δ}) as reported in (3).

$$TWR_{\delta} = \frac{TWD}{MRD} = \frac{MRT}{MRW \cdot \delta} \quad (3)$$

4 Results and Discussion

In this section, the distribution of energy would be discussed, analysing the efficiency in terms of specific removal energy. Figure 4a shows the ratio between TWD and the energy of single discharge (TWD_E) as a function of the reinforcement fraction and the pulse type. In the same way, the MRD_{δ} was related to the energy of a single discharge to evaluate the energy efficiency from the material removed point of view. In both cases (Fig. 4a, b), pulse type A shows a lower efficiency; in other words, this means that most of energy developed in a single discharge was lost. This is a positive effect from the tool wear point of view; in fact, low TWD_E means low tool wear per discharge, and consequently a lowest tool wear. In this case, the pulse type C is characterized by a higher fraction of energy dedicated to the material removal from the workpiece, despite pulse type C is being characterized

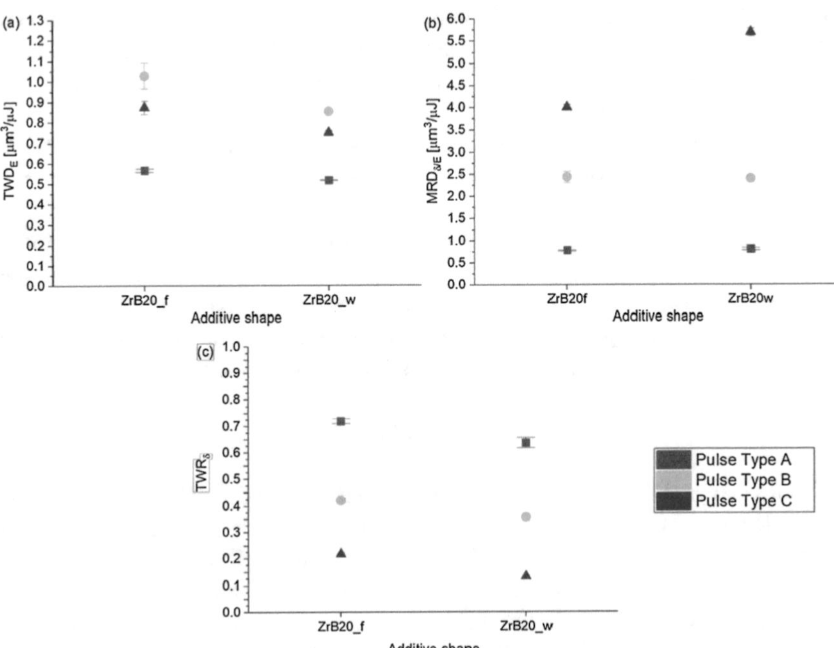

Fig. 4 Average values and standard deviation as a function of the reinforcement shape and pulse type, of **a** ratio between TWD and energy per discharge, estimated considering the relative density of the samples, **b** MRD and energy per discharge estimated considering the relative density of the samples, **c** TWR

Table 3 Analysis of variance *p*-values

		Factors		
		Pulse type	Reinforcement shape	Pulse type * reinforcement shape
Indicators	$MRD_δ$ (μm^3)	0.000	1.21×10^{-4}	0.030
	TWD (μm^3)	0.000	0.000	4.51×10^{-5}
	TWRδ (–)	0.000	0.000	0.202
	Sa (μm)	0.000	0.005	0.156

by a very short duration and low energy (Fig. 4b). Despite the great difference in energy per discharge, this shows that the pulse type C directs most of the energy towards the workpiece. Figure 4c shows that the TWR for all pulse types is lower for specimens. In particular, the whiskers show a better performance allowing to reduce the tool wear and to increase the material removal rate efficiency. The intermediate position of pulse type C identified in Fig. 4a can be related to a different energy distribution. In particular, which energy fraction is dedicated to the material removal from the workpiece. The factorial design was analysed in order to comprehend which factors and interactions are statistically significant for the performance indicators and surface roughness. A general linear model was used to perform a univariate analysis of variance, including all the main factors and their interactions. The ANOVA results of the experimental plan are reported in Table 3. The parameters are statistically significant for the process when the *p*-value is less than 0.05 (a confidence interval of 95% is applied). As a general remark, all the indicators resulted to be influenced by both the reinforcement shape and the pulse type. In some cases (for $MRD_δ$ and TWD), also the interaction showed an effect in terms of ANOVA. This aspect suggested that the interaction of factors is relevant for indicators that, in some way, can be correlated to the machining duration.

Main effects plots (Fig. 5) show that indicators are mainly influenced by the pulse type that establishes the range in which process parameters can vary, and in particular, the characteristics of the pulses. For all indicators, reduction in pulse duration and in peak current intensity generate the lower value of $MRD_δ$, TWD, and $TWR_δ$. At the same time, tests with whiskers reinforcement generate an improvement in $MRD_δ$ and in surface finishing while giving rise to a reduction in TWD and $TWR_δ$. By increasing the pulse duration and the peak intensity from type C to type A, the $MRD_δ$ is 10 times greater, but the surface quality decreases by −60%. For $MRD_δ$ and TWD, also the interaction between pulse type and reinforcement shape affects the results. In particular, when the samples are machined by pulse type C, the effect on $MRD_δ$ is more evident, while from the TWD point of view it is more evident for pulse type A (Fig. 6). In general, tests performed on materials with whiskers reinforcement are characterized by better results, both in terms of process performances and surface finishing. In fact, optimal performances for ED-machining are characterized by high level of $MRD_δ$, to perform fast machining, and low values of TWD, to reduce the waste of material related to the tool wear.

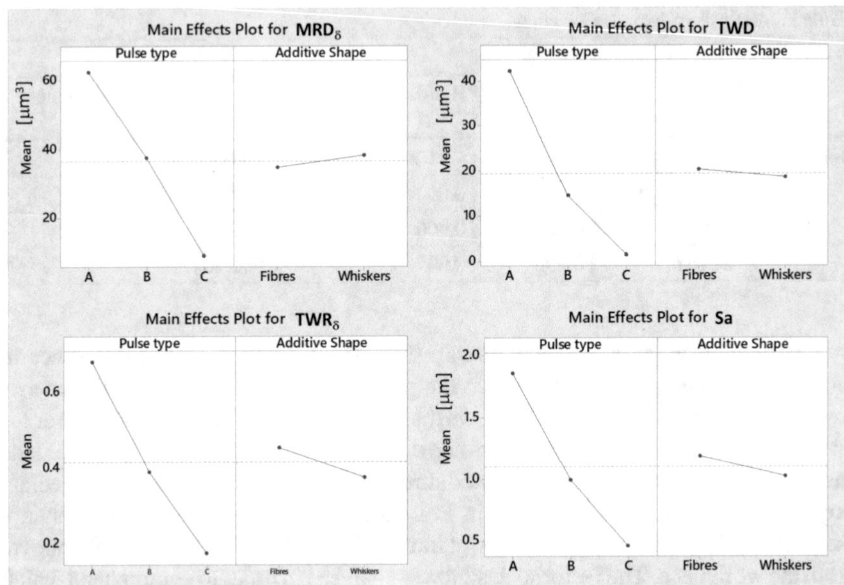

Fig. 5 Main effects plot for indicators affected by pulse type and reinforcement shape

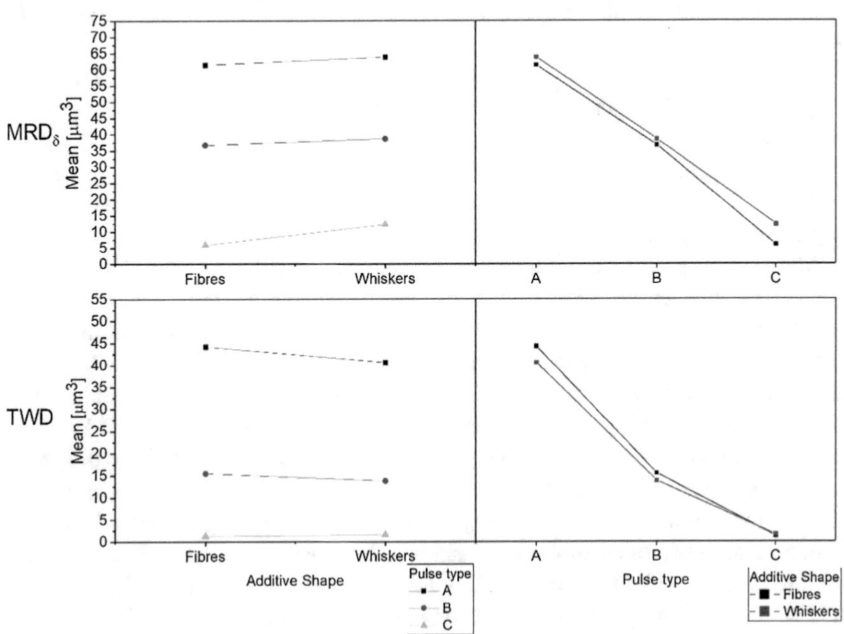

Fig. 6 Interaction plot for MRD$_\delta$ and TWD

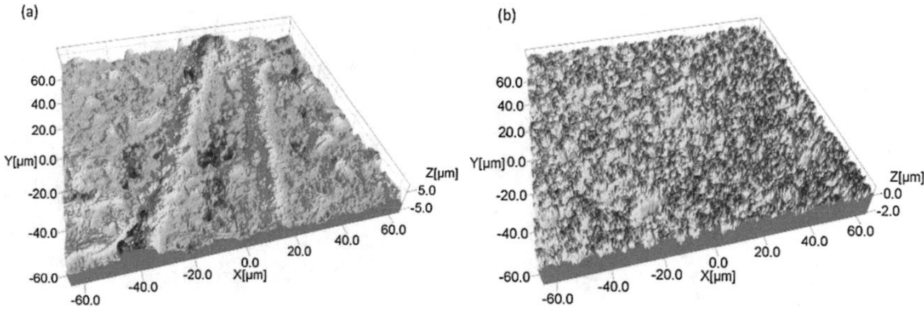

Fig. 7 Details of machined surfaces for ZrB20f (**a**) and ZrB20w (**b**) machined by pulse type A

A 3D reconstruction of an ED-machined surface detail scanned by a confocal laser scanning microscope with a magnification of 100× is reported as an example in Fig. 7. In particular, Fig. 7a represents a portion of the machined area on ZrB20f. Here it is possible to identify a sort of "protrusion" in correspondence of the fibres. This aspect is probably related to not complete machining because of the SiC low electrical conductivity characteristic and the great extension of the fibres area. Figure 7b represents the ED-machined surface for the sample containing SiC whiskers. In this case, the surface appears uniform and homogeneous because the SiC particles, as reported in the materials section, are better dispersed in the base matrix. These aspects justify the different results obtained in terms of surface quality and, in general, these different textures can be considered a starting point for further studies about the material removal mechanism occurred on UHTCs, in particular when there are low-electrically conductive parts in the structure. Form the 3D reconstruction it is possible to observe that the surfaces are not characterized by the typical aspect of the ED-machined surfaces which present a texture well-described by the presence of craters. In this specific case, the UHTC is different, but the same considerations can be done. A sort of craters can be observed by means of SEM (Fig. 8).

Machined surfaces on specimens doped with SiC whiskers are characterized by higher fragmentation of the recast layer. This aspect is particularly evident on surfaces machined by long pulses; in fact, for pulse type A and B, the surface is for the major part covered by recast materials for both reinforcement shapes, but the specimens doped with whiskers present smaller extensions of the single "crater" of recast materials.

A different behaviour can be observed for surfaces machined by the short pulses. In this case, for specimens containing the SiC fibres can be observed a smaller area of recast material, probably due to the combination of the greater dimensions of the fibres (in comparison to the whiskers), the lower energy per discharge for short pulses and the low electrical conductivity of the SiC. In particular, the fibres have a bigger surface and it needs to remove the entire parts. For this reason, the surfaces containing fibres presented in their correspondence a sort of protrusion.

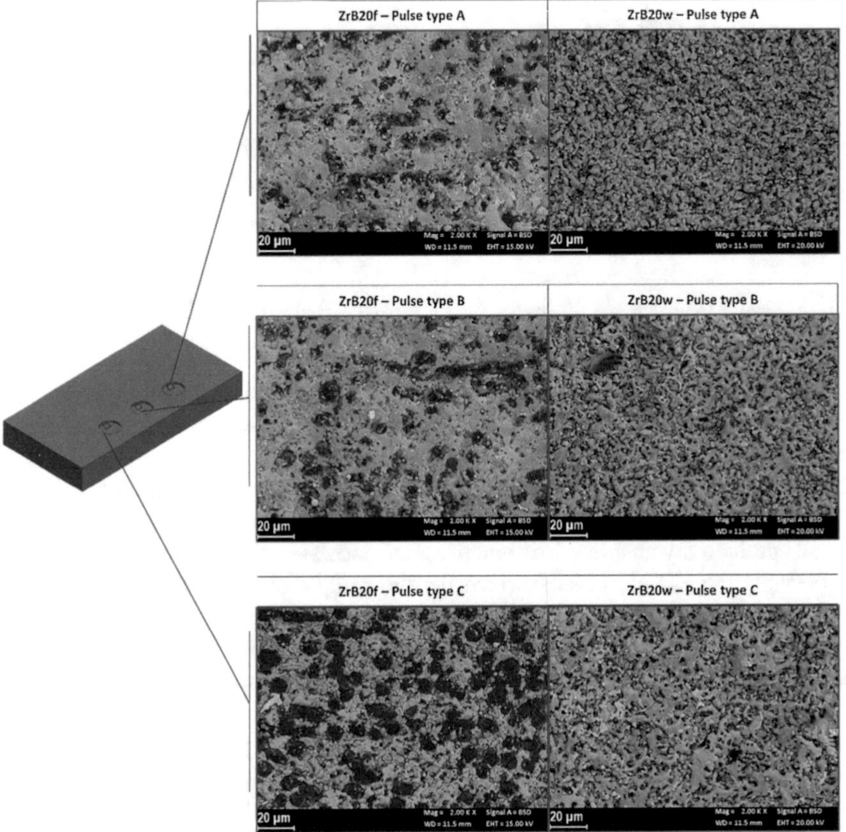

Fig. 8 SEM backscatter images of the machined surface

5 Conclusions

An evaluation of the machinability of ZrB_2-based composites hot-pressed with two different shapes non-reactive reinforcement (SiC) was performed in this work. Stability and repeatability of the μEDM were evaluated to identify the effects of the reinforcement shapes. The analysis took into account the process performances and surface finishing.

First of all, a discharge characterization was performed to feature the different pulse type used during the machining. In general, the discharge characterization and the performances indicators allowed to identify a stable and repeatable process with a faster material removal for samples doped with whiskers.

The analysis of variance showed that both factors, pulse type, and reinforcement shape, are statistically significant for the indicators selected in the process evaluation and in general, the use of whiskers improves the machining efficiency generating lower tool wear. Furthermore, the interaction between the two factors turns out to be influential only for MRD and TWD, which are indirectly related to the machining duration, since the number of discharges occurred during the machining was considered in their estimation.

This investigation shows that the specimens having a 20 vol.% of reinforcement in form of whiskers results to be the best solution in terms of machinability by EDM process not only for the better process performances (high MRD and low TWD), but also for the higher level of surface quality which is one of the essential criteria for making a proper decision in an industrial environment.

Acknowledgements The authors would like to thank D. Sciti and L. Silvestroni from ISTEC—CNR of Faenza (Italy) for the production and supply of the materials used in this study.

References

1. Saccone G, Gardi R, Alfano D, Ferrigno A, Del Vecchio A (2016) Laboratory, on-ground and in-flight investigation of ultra high temperature ceramic composite materials. Aerosp Sci Technol 58:490–497
2. Upadhya K, Yang JM, Hoffman W (1997) Advanced materials for ultrahigh temperature structural applications above 2000 °C. Am Ceram Soc Bull 58:51–56
3. Levine SR, Opila EJ, Halbig MC, Kiser JD, Singh M, Salem JA (2002) Evaluation of ultra-high temperature ceramics for aeropropulsion use. J Eur Ceram Soc 22:2757–2767
4. Justin JF, Jankowiak A (2011) Ultra high temperature ceramics: densification, properties and thermal stability. AerospaceLab J 3:AL3-08
5. Zhang P, Hu P, Zhang X, Han J, Meng S (2009) Processing and characterization of ZrB_2-SiCW ultra-high temperature ceramics. J Alloys Compd 472:358–362
6. Wang H, Wang CA, Yao X, Fang D (2007) Processing and mechanical properties of zirconium diboride-based ceramics prepared by spark plasma sintering. J Am Ceram Soc 90:1992–1997
7. Zhang X, Xu L, Du S, Liu C, Han J, Han W (2008) Spark plasma sintering and hot pressing of ZrB_2-SiCW ultra-high temperature ceramics. J Alloys Compd 466:241–245
8. Zhang X, Xu L, Han W, Weng L, Han J, Du S (2009) Microstructure and properties of silicon carbide whisker reinforced zirconium diboride ultra-high temperature ceramics. Solid State Sci 11:156–161
9. Yang F, Zhang X, Han J, Du S (2008) Processing and mechanical properties of short carbon fibers toughened zirconium diboride-based ceramics. Mater Des 29:1817–1820
10. Yang F, Zhang X, Han J, Du S (2009) Characterization of hot-pressed short carbon fiber reinforced ZrB_2-SiC ultra-high temperature ceramic composites. J Alloys Compd 472:395–399
11. Zhang X, Xu L, Du S, Han J, Hu P, Han W (2008) Fabrication and mechanical properties of ZrB_2-SiCw ceramic matrix composite. Mater Lett 62:1058–1060
12. Tian WB, Kan YM, Zhang GJ, Wang PL (2008) Effect of carbon nanotubes on the properties of ZrB_2-SiC ceramics. Mater Sci Eng A 487:568–573
13. Guicciardi S, Silvestroni L, Nygren M, Sciti D (2010) Microstructure and toughening mechanisms in spark plasma-sintered ZrB_2 ceramics reinforced by SiC whiskers or SiC-chopped fibers. J Am Ceram Soc 93:2384–2391
14. Silvestroni L, Sciti D, Melandri C, Guicciardi S (2010) Toughened ZrB_2-based ceramics through SiC whisker or SiC chopped fiber additions. J Eur Ceram Soc 30:2155–2164
15. Rezaie A, Fahrenholtz WG, Hilmas GE (2013) The effect of a graphite addition on oxidation of ZrB_2-SiC in air at 1500 °C. J Eur Ceram Soc 33:413–421

16. Hwang SS, Vasiliev AL, Padture NP (2007) Improved processing and oxidation-resistance of ZrB_2 ultra-high temperature ceramics containing SiC nanodispersoids. Mater Sci Eng A 464:216–224
17. Zhang X, Liu R, Zhang X, Zhu Y, Sun W, Xiong X (2016) Densification and ablation behavior of ZrB_2 ceramic with SiC and/or Fe additives fabricated at 1600 and 1800 °C. Ceram Int 42:17074–17080
18. Zhang L, Kurokawa K (2016) Effect of SiC addition on oxidation behavior of ZrB_2 at 1273 K and 1473 K. Oxid Met 85:311–320
19. Sciti D, Guicciardi S, Silvestroni L (2011) SiC chopped fibers reinforced ZrB_2: effect of the sintering aid. Scr Mater 64:769–772

Air Jet Cooling Applied to Wire Arc Additive Manufacturing: A Hybrid Numerical-Experimental Investigation

Filippo Montevecchi, William Hackenhaar, and Gianni Campatelli

Abstract WAAM (Wire Arc Additive Manufacturing) is a metal additive manufacturing process based on gas metal arc welding which enables to create large parts with a high deposition rate. WAAM is prone to the heat accumulation issue, i.e. a progressive increase of the workpiece internal energy due to the high heat input of the welding process, which can cause defects such as part structural collapse, uneven layers geometry or non-homogenous microstructure. A promising technique to mitigate such issue is to use an air jet impinging on the deposited material to increase the convective heat transfer. This paper presents an analysis of air jet impingement performances by means of a hybrid numerical-experimental approach. Different samples of a test case are manufactured using free convection cooling, air jet impingement and different interlayer idle times. Substrate temperatures are measured and compared with the results of a finite element simulation to assess its accuracy. The performances of air jet impingement are analyzed in terms of measured substrate temperatures and of simulated interlayer temperature, evaluated at the top of each layer. The results highlight that air jet impingement has a significant impact on the process, limiting the progressive increase of interlayer temperature compared with free convection cooling.

Keywords Wire-Arc-Additive-Manufacturing · Air cooling · Process FEM simulation

F. Montevecchi · G. Campatelli (✉)
Department of Industrial Engineering, University of Firenze, Via di Santa Marta 3, 50139 Firenze, Italy
e-mail: gianni.campatelli@unifi.it

F. Montevecchi
e-mail: filippo.montevecchi@unifi.it

W. Hackenhaar
Welding and Related Techniques Laboratory, Department of Mechanical Engineering, Federal University of Rio Grande do Sul, Porto Alegre, RS, Brazil
e-mail: william.hackenhaar@unifi.it

E. Ceretti and T. Tolio (eds.), *Selected Topics in Manufacturing*, Lecture Notes in Mechanical Engineering,
https://doi.org/10.1007/978-3-030-57729-2_6

Nomenclature

A, B, C	Empirical factors to calculate jet impingement heat transfer coefficient
l	Nozzle to target surface standoff distance
Nu	Nusselt number, i.e. dimensionless heat transfer coefficient
$\varnothing$	Angular coordinate for jet impingement heat transfer coefficient calculation
r	Radial coordinate for jet impingement heat transfer coefficient calculation
d	Air nozzle diameter
Re	Reynolds number of the impinging air jet
α	Angle between the jet axis and the target surface
Δy	Nozzle offset along the y coordinate
Δz	Nozzle to torch offset along the z coordinate

1 Introduction

WAAM (Wire Arc Additive Manufacturing) is an AM (Additive Manufacturing) process to produce metal components using direct deposition of arc welding beads. The advantages of WAAM, compared to other metal AM processes are [1]: (i) high deposition rate, (ii) possibility of manufacturing large parts, (iii) low capital investment. In WAAM, the high heat input of arc welding can lead to a progressive increase of the workpiece internal energy, known as heat accumulation [2]. This phenomenon occurs since the preferential cooling mode of the molten pool is the conduction towards the substrate. Increasing the number of deposited layers decreases the magnitude of the conductive heat flux, causing the heat accumulation issue. The consequences of this phenomenon are the increase of the molten pool size and of the interlayer temperature, i.e. the temperature of the top layer at the start of the subsequent one. Such effects result in modifications of both the layers geometry [3] and the material microstructure [4, 5] along the deposited height [6].

The most common way of preventing heat accumulation is to introduce interlayer idle times, i.e. allow the workpiece to cool down before depositing the subsequent layer [7, 8]. The drawback of this approach is that the idle times required to keep a constant interlayer temperature can be significantly high compared with the effective deposition time, resulting in a loss of productivity. An approach to decrease the interlayer idle times is to use active cooling systems to increase the heat extraction, i.e. the heat transfer to the environment. Takagi et al. [9] immersed the workpiece in a water cooled tank. Despite its effectiveness, this system is unpractical to be implemented on existing welding facilities (e.g. welding robot cells) which is one of WAAM main advantages. Li et al. [10] used a thermoelectric cooling system based on the Peltier effect to cool the side surfaces of a vertical wall, i.e. a workpiece made of straight layers. The proposed solution provided a significant reduction of bead unevenness, grain size and manufacturing time but its application to complex curved geometries is not straightforward. An alternative approach, easy to implement and suitable to manufacture complex parts, is to use a gas jet to increase the convective heat transfer coefficient. Wu et al. [11] used a CO_2 jet directed on the weld bead top surface by a nozzle attached to the welding torch. The system was tested depositing a Ti6Al4V vertical wall test case showing an improvement in material strength, hardness and manufacturing efficiency. A similar approach was also proposed by Montevecchi et al. [12], having two main differences with [11]: (i) the usage of standard compressed air rather than CO_2, which can reduce the operational cost for materials less sensitive to the welding atmosphere than Ti6Al4V, (ii) the different target of the jet, i.e. the already deposited layers over the current one, which enables to achieve jet impingement heat transfer, a highly efficient method to create a high thermal flux per unit surface [13]. The potential effectiveness of the proposed method was assessed [12] using a validated model of jet impingement heat transfer coupled with a validated thermal model of the WAAM process. The results highlighted that jet impingement prevented an excessive increase of both molten pool size and interlayer temperature, making it a promising approach to prevent heat accumulation.

In this paper, a cooling system based on air jet impingement was implemented on a prototype WAAM machine. A test case workpiece was selected, and different samples were manufactured using free convection and jet impingement cooling. The tests were carried out using different interlayer idle times to assess the performance of jet impingement cooling in different heat accumulation conditions. Substrate temperature was measured in different points using K-type thermocouples. The temperature data acquired during the deposition of the different samples were compared to assess the effect of air jet impingement on substrate temperature. In order to evaluate the effect of air cooling on interlayer temperature, the deposition of the samples was simulated using a FE (Finite Element) model. Simulations and thermocouple data were compared to evaluate the model accuracy. The interlayer temperatures were extracted to compare the pattern of air jet and free convection cooling samples.

2 Materials and Methods

To evaluate the performance of the air jet impingement, this paper uses a hybrid numerical-experimental approach. Different samples of the test case were manufactured measuring the substrate temperature with K-type thermocouples. However, substrate mounted thermocouples provide only a limited insight in the heat transfer phenomena occurring during the deposition, since they provide punctual data in a region far from the molten pool. To extend the investigation, a FE model was used to simulate the deposition of the test cases. Simulation results were then compared with thermocouples data to assess the accuracy of the model. Since the FE model returns the transient temperature field over the entire workpiece domain, an accurate prediction of substrate temperature enables to extend the heat transfer analysis to different regions. Therefore, the interlayer temperatures at the top of each layer were extracted from the simulation results, providing an important parameter to assess the effectiveness of jet impingement cooling. It must be pointed out that the presented analysis of the simulation results aims at providing a comparison of the interlayer temperature trend rather than accurately quantifying its punctual modifications.

Section 2.1 describes the experimental part of the work, i.e. the test case, the implementation of jet impingement cooling and the experiments conditions. Section 2.2 depicts the FE model used to simulate test case manufacturing.

2.1 Experiments Description

This paper aims at assessing the effectiveness of air jet impingement in preventing heat accumulation during GMAW (Gas Metal Arc Welding) based WAAM operations. The test case selected to perform the experiments is a vertical wall, whose geometry is presented in Fig. 1.

The wall height has a significant influence in determining the heat accumulation. The value of 50 mm was selected considering the test case geometry used by Wu et al. in [11]. The deposition was carried out using AWS ER70S-6, a standard filler metal for carbon steel welding. The substrate was a 30 mm thick AISI 1040 block, to prevent excessive distortions. The deposition was carried out using an AWELCO 250 Pulsemig GMAW welding unit connected to retrofitted 3-axes milling machine. The process parameters combination was selected to achieve a good bead appearance while maintaining a relatively low heat input: voltage 18 V (constant voltage); current 93 A; wire feed speed 4.6 m/min; nozzle to workpiece distance 10 mm; travelling speed 200 mm/min; heat input 4.95 kJ/mm; wire diameter 0.8 mm; shielding gas Ar + CO_2 17% supplied at 15 l/min. Such parameters resulted in welding beads 7.5 mm wide and 2.0 mm thick, requiring 25 layers to manufacture the selected geometry. It must be pointed out that the accuracy of the FE modelling techniques used in this paper was previously verified with this set of process parameters, strengthening the relevance of the simulation-based interlayer temperature analysis. The test case

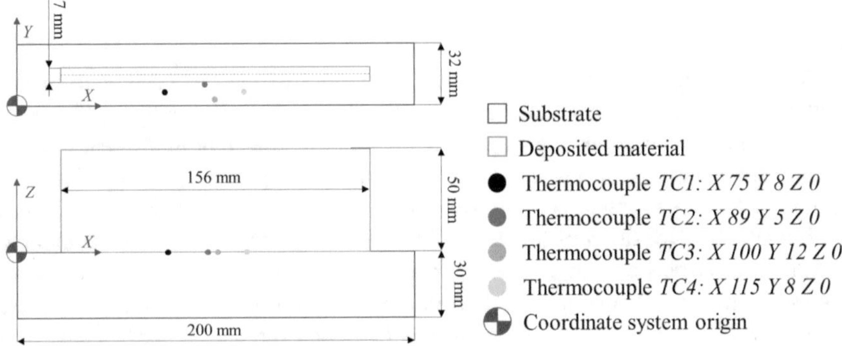

Fig. 1 Test case geometry and thermocouples layout

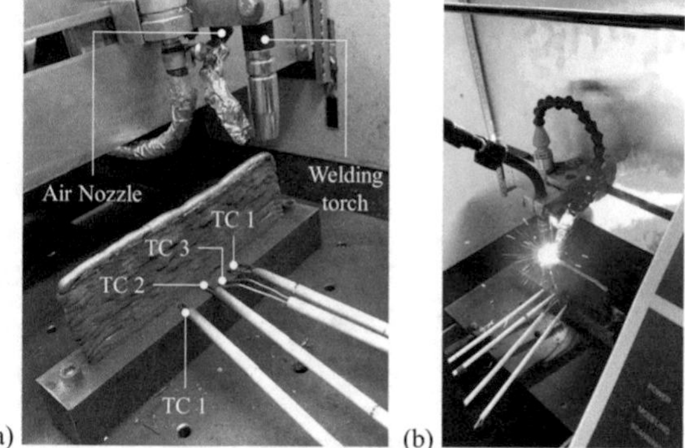

Fig. 2 **a** Proposed cooling system, thermocouples arrangement and as deposited test case; **b** test case deposition using air jet cooling

was positioned in the WAAM machine as shown in Fig. 2a. Figure 1 highlights the orientation of the workpiece with respect to the axes of the WAAM machine.

The temperature of the substrate was monitored using 4, 1.29 mm diameter, non-shielded, K-type thermocouples, with exposed hot junction and ceramic insulation. The thermocouples were fixed to the top surface of the substrate by spot welding, as shown in Fig. 2a. The positions of the thermocouples with respect to the coordinate system are marked in Fig. 1. Using multiple measurement points allowed to carry out a more detailed comparison between FE simulations and experiments. Thermocouples data were acquired using a National Instruments 9212 thermocouple module connected to a PC. All signals were recorded at a sampling rate of 10 Hz.

A coolant hose for machining applications with a 3 mm diameter nozzle was attached to the torch support. The low diameter nozzle was chosen to position it at different angles in the vicinity of the welding torch. The hose was connected to 0.2 m^3 plenum, supplied with dry air at the pressure of 0.6 MPa. The air flow was initiated by a solenoid valve controlled by the numerical control of the WAAM. Figure 2a shows the implementation of the cooling system in the WAAM machine while Fig. 2b shows the deposition of a sample carried out using the proposed air jet cooling system.

The selection of jet parameters is important to achieve a high convective heat flux. According to Goldstein and Franchett [14] the local heat transfer coefficient of an air jet impinging on a target surface can be calculated using the correlation depicted in Eq. (1):

$$Nu(r, \phi) = A * Re^{0.7} exp - \left(B + C \cos(\phi) \left(\frac{r}{d} \right)^{0.75} \right) \quad (1)$$

Nu is the Nusselt number, i.e. the dimensionless form of the heat transfer coefficient; *Re* is the Reynolds number evaluated at the nozzle outlet section; *r* and ø define the polar coordinates of target surface points in system centered in the intersection between the jet axis and the target surface; *A*, *B* and *C* are dimensionless coefficients determined by experiments. Goldstein and Franchett [14] highlighted

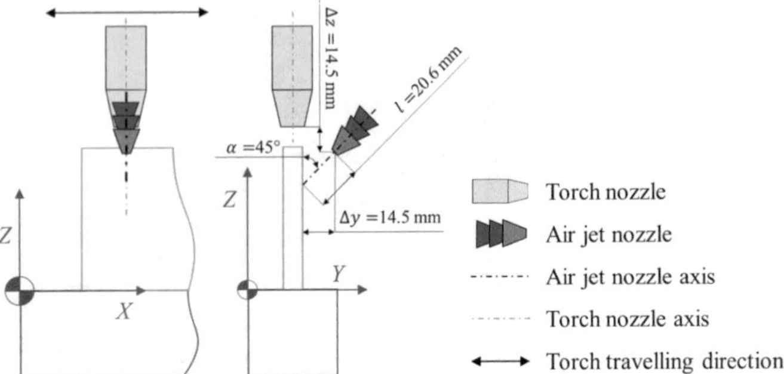

Fig. 3 Air jet arrangement in the experiments

that such coefficients depend on angle (α) between the jet axis with the target surface and on the ratio (l/d) between the nozzle standoff distance and the nozzle diameter. Equation (1) highlights that the heat transfer coefficient has an exponential decay from its maximum value at the intersection point. For decreasing α and increasing l/d values, Nu experiences an increase in the decay coefficient and in the peak value. Therefore, for a given nozzle diameter and air flow rate, a perpendicular jet and a low standoff distance increase the heat transfer coefficient. However, in WAAM, the increase of both α angle and nozzle proximity is limited by the interference of the air jet with the arc shielding gas, since an excessive mixing can lead to arc instabilities or promote the oxidation of the deposited material. Preliminary tests carried out with the nozzle perpendicular to the wall surface highlighted that this condition did not allow to achieve a stable arc, independently on the air flow rate and on the standoff distance. Based on these tests, the nozzle position and orientation depicted in Fig. 3 were adopted for the scope of this paper.

The axis of the air nozzle was located in the same XZ plane of the torch axis, allowing the jet axis to intersect the wall surface for every X position of the torch, i.e. during the deposition of the entire layer. The offsets Δz and Δy (highlighted in Fig. 3) define the position of the nozzle in the YZ plane. The offset Δy was set to the minimum possible value to avoid its interference with the substrate. For what concerns Δz, decreasing its value moves the intersection of the jet axis with the target surface closer to the molten pool. This results in a higher fluid to surface temperature gradient, i.e. in a more efficient heat extraction. However, an excessive vicinity to the mouth of the torch would result in an excessive contamination of the inert gas. Therefore, the 14.5 mm value (Fig. 3) was selected as a tradeoff between these opposite requirements. Based on an analysis of the Goldstein and Franchett correlation [14], the α angle was set to 45° since lower angles resulted in a detrimental decrease of the heat transfer coefficients. Such geometrical arrangement resulted in a l distance of 20.6 mm, i.e. in a l/d ratio of 6.9.

Further preliminary tests were carried out using the presented jet orientation and location to identify the maximum achievable flow rate to prevent arc instability issues.

Table 1 Summary of manufactured test cases

Test ID	Interlayer idle time (s)	Cooling condition	Initial jet cooling layer
1	120	Free convection	None
2	120	Jet impingement	14
3	30	Free convection	None
4	30	Jet impingement	11
5	10	Free convection	None
6	10	Jet impingement	11

The *Re* value of the selected flow rate was 22,000, which is close to the center of the *Re* range covered by the Goldstein and Franchett [14] correlation.

As earlier mentioned, different samples of the test cases were manufactured using different idle times both in free convection and air jet cooling conditions. For all the samples, the first 10 layers were deposited using standard cooling and 120 s of interlayer idle time. This strategy was selected since activating the jet cooling below this level did not allow to target the wall surface with the air jet. Table 1 summarizes the conditions of each test, namely the idle times, the cooling conditions and the layer in which the air cooling is activated.

2.2 Finite Element Modelling

The deposition process was simulated using a FE heat transfer analysis. The simulations were carried out using the commercial FE solver LS-DYNA. The WAAM

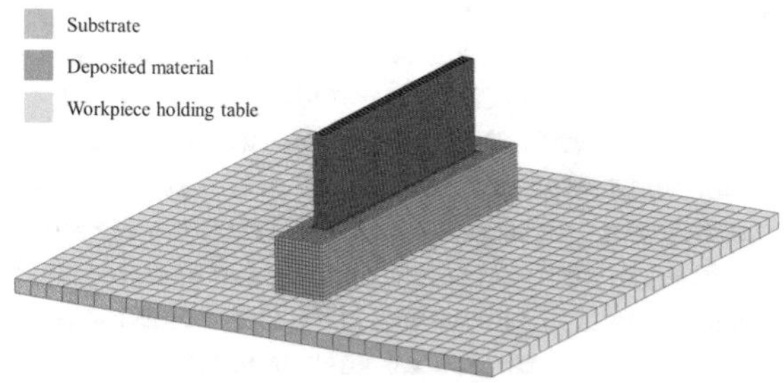

Fig. 4 FE model of the test case

modelling techniques, such as the moving heat source, the elements activation algorithm and the material behavior models are based on literature works [15, 16]. Figure 4 shows the FE model used to simulate the deposition of the test cases.

The geometry is discretized using 60,274 1st order solid hexahedral elements, resulting in 75,017 nodes. The modelling domain was extended to the mild steel workpiece holding table since the conductive heat flux from the substrate is a relevant contribution to the overall heat extraction. The conduction between the substrate and the workpiece holding table was included using a contact algorithm. The interface conductance was set to 2000 W/m^2 °C based on literature data [17]. The effect of air jet impingement was included using the technique presented in [12]. The different idle times were simulated by varying the length of the heat source on-off intervals.

3 Results and Discussion

This section discusses the results of the hybrid numerical experimental investigation of air jet impingement effectiveness. All the manufactured samples did not show evidence of any visible discontinuities. Section 3.1 presents the experimental results, discussing the overall trends of the thermocouples in the different tested conditions. Section 3.2 presents the comparison between the thermocouples and the simulation data, extending the comparison between air jet impingement and free convection cooling to the interlayer temperature pattern.

3.1 Thermocouple Results

This section discusses the results of the substrate temperature measurements. As earlier mentioned, an initial set of layers was deposited using an interlayer idle time of 120 s. Such value was then reduced according to Table 1. This section presents the results of the thermocouple TC2, since it was the closer to the deposited material. Figure 5 shows the results of experiments 1, 3, and 5, i.e. the depositions carried out without air jet cooling.

For all the tests, temperature curves show the typical cyclic pattern, due to the repeated passages of the welding torch. However, substrate temperatures show different patterns as the idle time is modified. The idle time of 120 s shows a slightly decreasing temperature trend per each cycle, i.e. both peak and minimum values experience a progressive decrease. These results are in accordance with the effect of idle times in thermal cycles found by Lei et al. [8] and the decrease of overall temperature as layers are deposited [18]. This behavior is due to the increase in deposited metal mass and surface for each deposited layer, which respectively increase the overall heat capacity and the heat extraction by convection and radiation. Therefore, longer idle times allow these factors to dominate the heat transfer problem over the reduction of the conductive heat flux due to the increase of workpiece height.

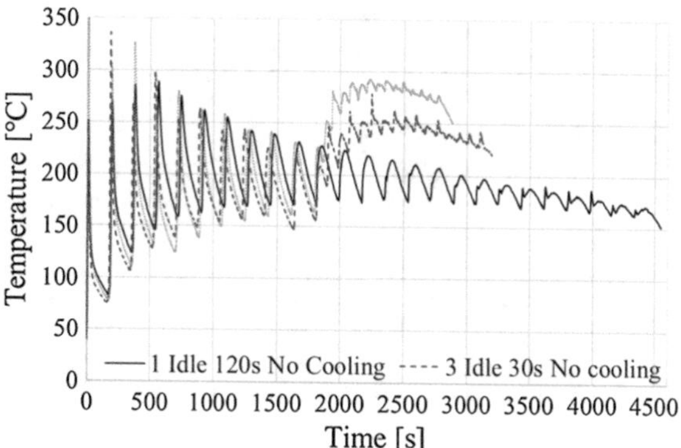

Fig. 5 Deposited layers with no cooling. 1, 120 s. 3, 30 s. 5, 10 s

However, the adoption of longer idle times has the negative effect of decreasing the productivity. For this reason, lower idle times of 30 and 10 s were tested, as shown in Fig. 5. Both tests show a significant increase of the average temperature after the decrease of the idle times. This trend of substrate temperature is expected to result in a significant increase of the interlayer temperature in the top layers. These hotter regions can cause negative effects on layers geometry, surface quality and material properties.

The experiments with air jet cooling were performed using the same idle times schemes used in the experiments without cooling. As mentioned in Sect. 2, the first set of layers was deposited with 120 s of interlayer idle time and no cooling.

Figure 6 compares the results of the experiments using air jet cooling, i.e. tests 2; 4 and 6. It is highlighted that in the test carried out using air jet impingement and 10 s idle time, the trendline lays below 250 °C, unlike in the same test performed using free convection cooling. This is clearly highlighted by Fig. 7b, which compares the results of tests 3 and 6.

Figure 7b highlights also that, despite the usage of cooling, using an idle time of 10 s does not prevent the sudden increase in substrate temperature. However, Fig. 7a, which compares the results with and without air cooling using 120 s idle

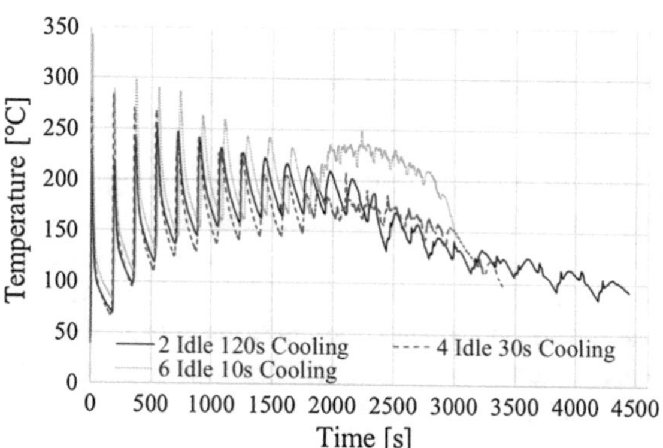

Fig. 6 Idle times for cooled deposits. 2, 120 s. 3, 30 s. 6, 10 s

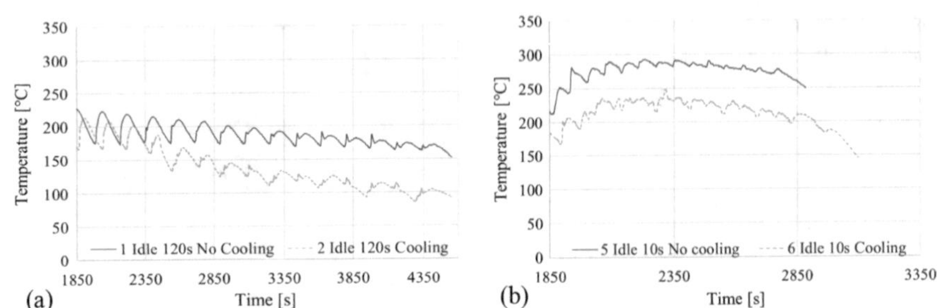

Fig. 7 Comparison of standard cooling and air jet cooling: **a** 120 s idle time; **b** 10 s idle time

time, highlights that in this case the usage of cooling results in a steeper decrease of the average substrate temperature. A similar trend occurs for 30 s idle (Fig. 6, test 4), where the substrate temperature experiences a significant reduction after the activation of the air jet. Moreover, besides the similarities in the trend, tests 2 and 4 show close punctual values after the activation of the air jet. This is an interesting result since, unlike in the free convection cooling tests, when using jet impingement, increasing the interlayer idle time above 30 s does not produce any significant reduction of the substrate temperature. Therefore, increasing the idle time would result only in a loss of productivity. For the specific test case, reducing the idle time from 120 to 30 s would reduce of 75% the WAAM machine inactive times during the deposition of the last layers. Considering the cost compared to industrial gases, the use of compressed air in deposited beads through an air nozzle presents as a good solution, since it would not require any relevant additional cost to achieve an increase in productivity.

However, considering the results of test 6, it is worth to remark that idle times cannot be reduced arbitrarily when using jet impingement. A limit idle time condition indeed exits for a given combination of process parameters, workpiece geometry, workpiece material and jet parameters. Reducing the idle times below such limit causes an insufficient heat extraction by the jet impingement cooling, which cannot prevent the excessive increase of the substrate temperature.

3.2 FEM Simulation Results

This section shows the effect of air jet impingement cooling by analyzing the results of FE simulations. The presented results are related to the simulations of test 1 and 2, i.e. using an interlayer idle time value of 120 s. Figure 8 presents the comparison of simulations results with thermocouples TC2 and TC4, since such thermocouples are respectively the closest and the furthest from the wall. The comparison was carried out by extracting the temperature time history of the nodes located in the closest position with respect to the thermocouples.

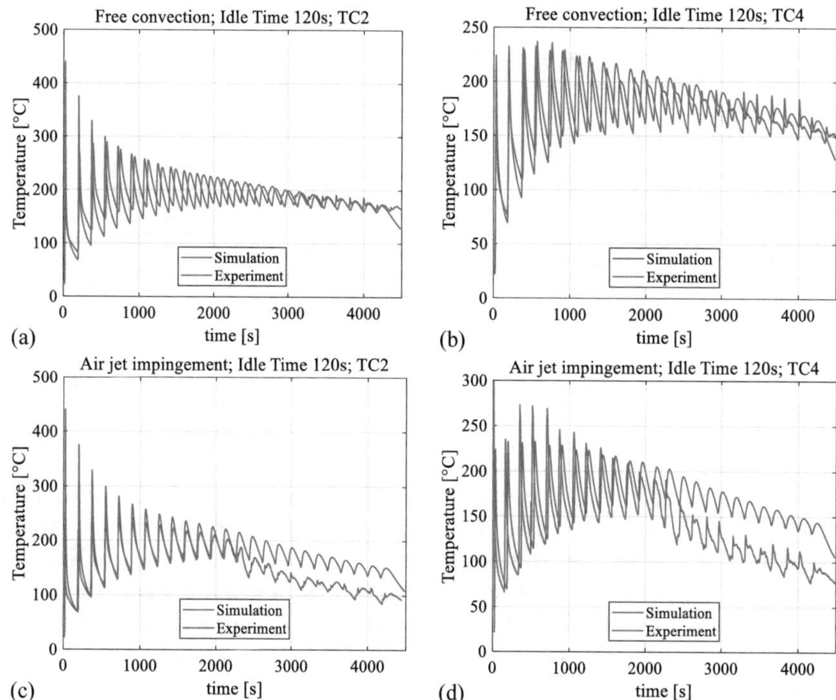

Fig. 8 Comparison of simulation and experimental results for the 120 s idle time tests: **a** thermocouple 2, free convection cooling; **b** thermocouple 4, free convection cooling; **c** thermocouple 2 air jet impingement; **d** thermocouple 4 air jet impingement

Figure 8a, b present the data related to test 1, i.e. in free convection cooling condition. The comparison highlights a general agreement between the measured and simulated temperature patterns. The FE model is indeed capable of predicting the trend of both minimum temperature per cycle and of peak to valley amplitude. For TC2, the FE model significantly overestimates first temperature peak. The reason for this overshoot could be related to inherent uncertainties in the positioning of the thermocouples. This has a significant impact since TC2 is the closest to the molten pool and during the deposition of the first layer it is located in a steep temperature gradient area.

Figure 8c, d present the results of the comparison for test 2, i.e. with air jet impingement. As for test 1, presented data highlights that the FE model predicts the overall trend of the temperature curves, including the effect of air jet impingement cooling. However, FE model is not in accordance with the experimental data for what concerns the decrease rate of the average temperature per cycle after the activation of jet impingement. The reason for this inaccuracy could be an underestimation of jet impingement heat transfer coefficients. The correlation of Goldstein and Franchett was indeed developed studying air jets impinging on a smooth surface. The air jet

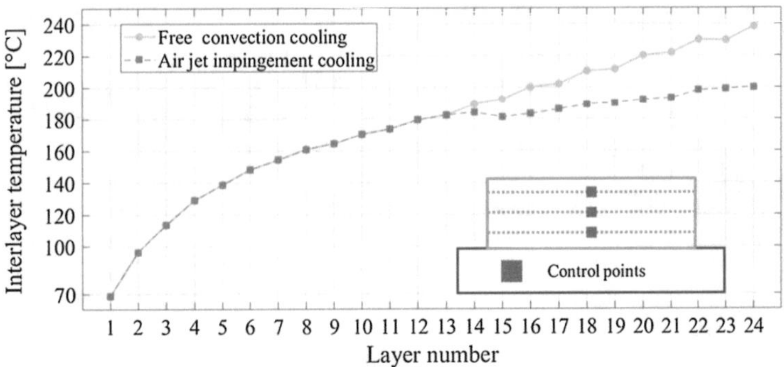

Fig. 9 Comparison of the simulated interlayer temperatures for the free convection and the air jet impingement cooling conditions

impingement boundary condition includes only the target surface of the vertical wall. However, the air flow is deflected also on the top surface of the substrate, in which the convective heat transfer is modelled as free convection since it is difficult to quantify the correct heat transfer coefficient. This leads to a lower heat extraction than in the experiments. Moreover, WAAM workpieces have a significant surface waviness which could increase the flow turbulence and consequently the heat transfer coefficients.

Despite the inaccuracies, the simulations showed a good agreement with the experimental data regarding the overall temperature trends, making them suitable for an analysis of the effect of air jet impingement on the interlayer temperature pattern. A set of control points was defined, located at the central point of the top surface of each deposited layer. For each control point, the first local minima in temperature time history was considered as the interlayer temperature, since it is the minimum temperature after the deposition of the layer associated with the current control point. Figure 9 presents the results of the interlayer temperature comparison and the control points location.

In both curves the interlayer temperature increases throughout all the deposition process. However, the curve related to the air jet impingement simulation show a significant change in the curve slope after the activation of the air cooling. This confirms that the reduction of the substrate temperature observed in the thermocouple data indicates that jet impingement can limit the uncontrolled increase of the interlayer temperature. Moreover, considering the results presented in Fig. 8c, d, jet impingement effect could be underestimated in the FE simulation. This could result in a further decrease of the slope of the interlayer temperature curve using air jet impingement.

4 Conclusions

This paper presents an analysis of the effectiveness of air jet impingement as a cost effective and versatile active cooling strategy to prevent heat accumulation in WAAM. A test case was manufactured using different interlayer idle times. Different samples were manufactured both using traditional free convection and air jet impingement cooling. Jet impingement effectiveness was assessed using a hybrid numerical-experimental method, in which measured substrate temperatures were compared to FE simulations results. The verified simulation data were then used to extract the interlayer temperature during the deposition process.

The presented data highlighted that the air jet impingement enabled to reduce the substrate temperature in all the tested configurations. In the tests conditions, increasing the idle time from 30 to 120 s when using air jet impingement did not result in a further decrease of the substrate temperature. However, the tests carried out using 10 s idle time highlighted that in this condition, air jet impingement could not prevent an increase in the substrate temperature but could only reduce its magnitude. This suggests that, for a given process and jet conditions, there is a limit to the effectiveness of air-cooling.

Despite some inaccuracies, the FE simulations accurately predicted the overall trend of substrate temperature. The analysis of the results highlighted that the decreasing trend of the substrate temperature, resulted in a significantly slower increase of the interlayer temperature. This confirmed that air jet impingement can be an effective approach to control the interlayer temperature without a relevant increase of the idle times.

The main outlook of the presented research is to investigate the effect of air jet impingement on the properties and microstructure of materials sensitive to heat accumulation, such as Inconel 625, stainless steels or Ti6Al4V. Different idle times and jet parameters should be tested to assess if, under specific conditions, jet impingement cooling could have a positive effect on the deposited material properties and microstructure. This analysis could be enhanced by increasing the accuracy of the FE model using air jet impingement. This would enable to predict the jet parameters and the idle times to achieve a target interlayer temperature, for a given set of workpiece geometry, material and process parameters.

References

1. Cunningham CR, Flynn JM, Shokrani A, Dhokia V, Newman ST (2018) Invited review article: strategies and processes for high quality wire arc additive manufacturing. Addit Manuf 22:672–686
2. Wu B, Ding D, Pan Z et al (2017) Effects of heat accumulation on the arc characteristics and metal transfer behavior in wire arc additive manufacturing of Ti6Al4V. J Mater Process Technol 250:304–312

3. Xiong J, Li Y, Li R, Yin Z (2018) Influences of process parameters on surface roughness of multi-layer single-pass thin-walled parts in GMAW-based additive manufacturing. J Mater Process Technol. 252:128–136
4. Wu B, Pan Z, Ding D, Cuiuri D, Li H (2018) Effects of heat accumulation on microstructure and mechanical properties of Ti6Al4V alloy deposited by wire arc additive manufacturing. Addit Manuf 23:151–160
5. Wang JF, Sun QJ, Wang H, Liu JP, Feng JC (2016) Effect of location on microstructure and mechanical properties of additive layer manufactured Inconel 625 using gas tungsten arc welding. Mater Sci Eng A 676:395–405
6. Prado-Cerqueira JL, Diéguez JL, Camacho AM (2017) Preliminary development of a wire and arc additive manufacturing system (WAAM). Procedia Manuf 13:895–902
7. Montevecchi F, Venturini G, Grossi N, Scippa A, Campatelli G (2018a) Idle time selection for wire-arc additive manufacturing: a finite element-based technique. Addit Manuf 21:479–486
8. Lei Y, Xiong J, Li R (2018) Effect of inter layer idle time on thermal behavior for multi-layer single-pass thin-walled parts in GMAW-based additive manufacturing. Int J Adv Manuf Technol 96(1–4):1355–1365
9. Takagi H, Abe T, Cui P, Sasahara H (2015) Mechanical properties evaluation of metal components repaired by direct metal lamination. Key Eng Mater 656–657:440–445
10. Li F, Chen S, Shi J, Zhao Y, Tian H (2018) Thermoelectric cooling-aided bead geometry regulation in wire and arc-based additive manufacturing of thin-walled structures. Appl Sci 8(2):207
11. Wu B, Pan Z, Ding D, Cuiuri D, Li H, Fei Z (2018) The effects of forced interpass cooling on the material properties of wire arc additively manufactured Ti6Al4V alloy. J Mater Process Technol 258:97–105
12. Montevecchi F, Venturini G, Grossi N, Scippa A, Campatelli G (2018b) Heat accumulation prevention in wire-arc-additive-manufacturing using air jet impingement. Manuf Lett 17:14–18
13. O'Donovan TS, Murray DB (2007) Jet impingement heat transfer—part I: mean and root-mean-square heat transfer and velocity distributions. Int J Heat Mass Transf 50(17–18):3291–3301
14. Goldstein RJ, Franchett ME (1988) Heat transfer from a flat surface to an oblique impinging jet. J Heat Transfer 110(1):84
15. Montevecchi F, Venturini G, Scippa A, Campatelli G (2016) Finite element modelling of wire-arc-additive-manufacturing process. Procedia CIRP 55:109–114
16. Montevecchi F, Venturini G, Grossi N, Scippa A, Campatelli G (2017) Finite element mesh coarsening for effective distortion prediction in wire arc additive manufacturing. Addit Manuf 18:145–155
17. Sridhar MR, Yovanovich MM (1996) Thermal contact conductance of tool steel and comparison with model. Int J Heat Mass Transf 39(4):831–839
18. Lundbäck A, Lindgren LE (2011) Modelling of metal deposition. Finite Elem Anal Des 47(10):1169–1177

Evaluation of the Shear Properties of Long and Short Fiber Composites Using State-of-the Art Characterization Techniques

Antonios G. Stamopoulos, Alfonso Paoletti, and Antoniomaria Di Ilio

Abstract As the use of short fiber and textile thermoplastic composites is expanding in many industrial fields, particularly the automotive, it is necessary for each manufacturer in every sector to assess the mechanical characteristics and behavior of these materials in various loading and environmental conditions. Among the most difficult mechanical tests are those for calculating the shear properties and behavior of these materials. As a result, a variety of standards have been developed throughout the years. Among these, the most promising one may be considered the V-notched Rail Shear test as it incorporates the unique features of two different mechanical tests namely the Iosipescu and rail shear test. In the present work, this state-of-the-art mechanical test, originally designed for unidirectional composites, was implemented in different material architectures and its apparatus was modified and used for mainly two purposes; i.e. the investigation of the effect of the infusion direction on the mechanical properties of short and the warp-weft direction of the woven textile thermoplastic composites, as well as producing reliable data regarding the mechanical characteristics and the behavior of these materials for simulating the manufacturing process. The testing device was developed in a way to produce robust and accurate results, ensuring the alignment of its components as well as the stress uniformity in the section gauge. The results revealed a dependency of the infusion direction of the short fiber thermoplastics and the textile direction of the woven composites. To this end, this mechanical testing technique may be considered as a benchmark on the material characterization in shear deformation.

A. G. Stamopoulos (✉) · A. Paoletti · A. Di Ilio
Department of Industrial and Information Engineering and Economics (DIIIE), University of L'Aquila, Monteluco di Roio, 67100 L'Aquila, Italy
e-mail: antonios.stamopoulos@univaq.it

A. Paoletti
e-mail: alfonso.paoletti@univaq.it

A. Di Ilio
e-mail: antoniomaria.diilio@univaq.it

E. Ceretti and T. Tolio (eds.), *Selected Topics in Manufacturing*,
Lecture Notes in Mechanical Engineering,
https://doi.org/10.1007/978-3-030-57729-2_7

Keywords Thermoplastic composites · Mechanical testing · Shear characterization

1 Introduction

In the recent years, considerable attempts were conducted by all the sectors of industry towards the substitution of conventional with more innovative, high performance and more environmental-friendly materials. More particularly, in both the aeronautic and the automotive industry, a huge effort has been given to reduce the weight of the vehicles and to contribute this way to the reduction of the CO_2 emissions. To this end, a notable observation was the fact that the increment of the use of composite materials in the automotive industry helped the automotive companies to cope with the emissions' standardization in Europe and in Japan [1, 2]. Starting with the aeronautics industry, new aircrafts were produced that comprise more than 50% of composite materials, such as the Airbus A350 (53%) and the Boeing 787 (50%). The aeronautical manufacturers took into consideration the unique features of composites into the design process but there are several defects of these materials related mostly with the manufacturing process [3]. Most of these defects (fiber misalignment, porosity) decrease the properties of these materials, especially those related to the matrix properties (shear, flexural) [4, 5]. On the other part, the tendency of car manufacturers was well addressed by Mangino et al. [6], indicating also the skepticism towards the complete substitution of the conventional materials with composites. It is also underlined the necessity of the composite parts manufacturers to achieve high production component volumes which may easily be repaired, formed and recycled. Moreover, the automotive companies are using incrementally the new materials and material manufacturing technologies to new, more sophisticated vehicles which exploit optimally the benefits of the lightweight alloys and composites [7]. A good example is the implementation of carbon fiber reinforced materials in the chassis of the BMW i3 that allowed the manufacturers to use larger batteries (240 kg more). There are some predictions that in the future, a concept car may weight 40% less and may use composites as the primary material (more than 40%) [8].

To this end, the interest of the automotive companies was mainly concentrated to carbon or glass fiber reinforced thermoplastics (GFRTPs) mainly due to their recyclability and the good mechanical to weight properties. In addition, the use of short fiber reinforced thermoplastics has gained an increasing use in the sector of non-critical structural automotive parts as they exhibit satisfying stiffness and they can be produced via injection molding in high production rates reducing, at the same time, the fabrication cost [9]. In more demanding structures, in terms of loads and environmental conditions, the woven textile glass fiber reinforced thermoplastics tend to be utilized.

Nevertheless, as these materials still exhibit a certain anisotropy, a crucial information may concern their mechanical performance, mostly related to the matrix properties, under several loads and conditions. It is well known that several manufacturing defects contribute to the decrease of the shear properties of composites.

For instance, it is widely accepted that the parameters of the injection molding of short fiber reinforced thermoplastics has a strong influence on the fiber alignment towards the injection direction [10–12] or even the population of the remained intact fibers [13, 14], influencing at the same time the mechanical properties of the component. When using textile composites, the mechanical properties are strongly related with the fiber direction (warp or weft) and different results may be obtained while testing them as seen in [15]. Thus, the result of a manufacturing process can only be addressed via mechanical testing of specimens which are representative of the whole component. Consequently, the precise evaluation of the shear properties is based mostly on the representativeness of the specimens and the accuracy of the existing mechanical testing methods. In addition, via mechanical testing can be assessed not only the adequacy of the material in terms of structural condition but also the adequacy of a manufacturing process for the production of a component.

Among the most difficult mechanical tests for composite materials are those concerning the determination of the shear mechanical behavior and properties, especially those concerning the in-plane. Throughout the years, a number of standardized test methods have been proposed. Among these, the most popular are the Iosipescu, the Rail Shear and more recently the V-Notched Rail Shear test method. In the present work, this particular test method was considered as the base for the development of a modified testing apparatus for assessing the shear properties of composite materials of various types. To this end, the modified V-Notched Rail Shear apparatus was used for evaluating the shear response of short-fiber and textile fiber reinforced thermoplastics. During each test, the local and global deformations were measured and the behavior of each specimen under shear deformation was observed. The results of these mechanical tests revealed a strong influence of the direction of the injection molding to the shear properties of short fiber composites and a substantial independency from the warp and weft direction of the textiles. Considering the fact that the V-Notched Rail Shear testing method was developed for unidirectional (UD) composites, the present work contributes to the amelioration of this testing method in two main directions:

- modification of the apparatus for maintaining the specimen alignment during the test with less friction introduced possible;
- implementation of the testing method not only to UD composites but also to short fiber and textiles.

Moreover, other novelties of the present work are:

- the assessment of the effect of the injection direction to the in-plane properties of short glass fiber thermoplastics using the modified V-Notched Rail Shear apparatus;
- the investigation of the potential difference of the shear properties of textile glass fiber thermoplastics when loaded towards the warp or the weft direction.

2 V-Notched Rail Shear Testing Method

2.1 The State-of-the-Art Shear Mechanical Testing Methods

As previously mentioned, there have been developed at least 8 standardized mechanical tests for assessing the shear behavior of composite laminates. These standardized testing methods exhibit pros and cons which are synthetized in Table 1 [16]. As can be seen, the majority of these methods have significant drawbacks, especially concerning the shear stress uniformity and their flexibility for assessing all the shear strength characteristics of composite laminates. A typical example is the ±45° shear test which consists of a simple tensile specimen of a textile composite material. Although the profound simplicity of this test, the lack of stress uniformity and the fact that it can only be utilized for measuring the in-plane shear properties of textiles makes it inadequate for unidirectional or short fiber composites.

In addition to the standardized methods there have been developed also other non-standardized methods to overcome the difficulties of the previous ones concerning the complexity of the apparatus, the lack of features and the inaccuracy of the results.

More recently, standardized was a new testing apparatus and method call "V-Notched Rail Shear" [17, 18]. Its name is derived from the combination of two previously existing testing methods, the Iosipescu (ASTM D5379) [19] and the Rail

Table 1 The existing standardized testing methods for evaluating the shear properties of composite materials as presented in [16]

Test method	Standard	Features			
		Uniform shear stress	All 3 shear stresses practical	Shear strength obtained	Shear stiffness obtained
Short beam shear	ASTM D2344M			●	
Iosipescu shear	ASTM D5379	●	●	●	●
±45° shear	ASTM D3518			●	●
Two rail shear	ASTM D4255			●	●
Three rail shear	ASTM D4255			●	●
Double-notched shear	ASTM D3846			●	●
Thin tube torsion	ASTM D5448	●		●	●
V-notched rail shear	ASTM D7078	●	●	●	●

shear test (ASTM D4255) [20]. The specimen of this testing method is similar to the Iosipescu test, it consists of a rectangular specimen with a pair of v-shaped notches at the opposite sides of the specimen. This way the stresses are concentrated near the notches and the cross-sectional region between them is subjected to almost pure shear. The main drawback of the Iosipescu testing method comes from the way the load is transmitted from the apparatus to the specimen, a way which can easily introduce bending moments to the specimen. To overcome this problem, the specimen of the V-Notched Rail Shear method is wider compared to the Iosipescu and the load is applied via shear, just as in the Rail Shear test. Consequently, the ASTM D7078 [21] specimen is less prone to bending deformation and the load is applied uniformly to the specimen.

As the ASTM D7078 was recently introduced to the scientific world, the frequency of its utilization is still rare due to the complexity of the basic configuration of the apparatus. Nevertheless, in recent years it has gained an increasing attention as it overcomes the most significant drawbacks of the other methods. A very comprehensive comparison between the Iosipescu and the V-Notched Rail Shear methods was conducted by Almeida et al. [22] on glass fiber-epoxy composites where the superiority of the ASTM D7078 is underlined. This state-of-the-art method was successfully used for the identification and the evaluation of porosity effect on the shear behavior of UD carbon fiber reinforced epoxies [23]. In addition, this testing method was assessed numerically and experimentally by Taheri-Behrooz and Moghaddam [24] analyzing the behavior of glass/epoxy composites. It is underlined that the most common drawback of this method is the misalignment of the two parts of the apparatus and the misalignment of the specimen itself inside the tabs.

Taking into account the drawbacks of this testing method, Gude et al. [25] proposed two modifications, one regarding the relative alignment of the two main pieces of the apparatus and one regarding the clamping of the specimens for maintaining their proper alignment. In parallel this work represents a first attempt to implement this new testing method to textile composites. Nevertheless, the authors underline that this modification leaded to excessive friction load between the guide columns and the main part of the apparatus. Moreover, while the size of the fixture makes it rigid its weight is increased compared to the conventional ASTM D7078 [21] apparatus.

2.2 Design and Realization of the Modified V-Notched Rail Shear Apparatus

In the present work, the design of the testing apparatus was based on the following principles:

- rigidity of the apparatus;

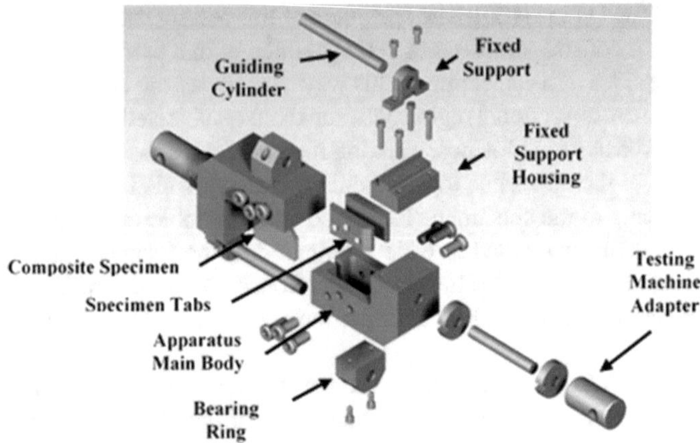

Fig. 1 Exploded view of the modified ASTM D7078 apparatus

- symmetry of the twin main pieces of the main body of the apparatus throughout the execution of a test so as to ensure the pure shear load and the corresponding load uniformity;
- simple to construct and to assemble (components simplicity). This way the apparatus may be constructed in every laboratory and machine shop;
- universality of its applicability in every universal testing machine (UTM);
- the designed testing apparatus should be able to test a variety of materials with different thicknesses varying from 0.5 to 5 mm.

Considering the specifications above, the designed modified V-Notched Rail Shear apparatus is presented in Fig. 1. As shown there, in the original apparatus proposed by the ASTM D7078 [21], 2 guiding cylinders were added in order to maintain the alignment of the two main pieces of the apparatus. To achieve the minimum friction possible, two linear ball guide were added to the columns. Another advantage of the use of these guides is the elimination of the spacers described by the standard ASTM D7078 which appear to be necessary while placing the specimen to the apparatus tabs. The material used for the main body is C45 structural steel, the self-aligning bearing rings and the fixed supports are catalogue commercial items from SKF® model LUND-12. This particular type of bearing rings may support static and dynamic perpendicular load up to 510 N and 695 N respectively [26]. The choice of this particular bearing rings was made on the base of the low friction values as indicated by the manufacturer (SKF®).

3 Experimental

3.1 Materials and Preparation

In the present work, the modified V-Notched Rail Shear testing method was applied to GFRTPs of two categories both with polypropylene matrix reinforced with short glass fibers and textile glass fibers. They both refer to the two most frequently used composites for manufacturing components in the automotive industry.

The polypropylene based PP-GF-30 (30% glass fiber content) short fiber reinforced material was firstly considered. The specimens for the execution of the mechanical tests were cut using water-jet cutting from PP-GF-30 plates made by injection molding. As mentioned in previous works [10–12], the fiber direction is a lot influenced by the injection direction, however the residual properties of the composite are very uncertain, since the alignment of the fibers is not regular. Thus, the specimens were cut in two directions, one parallel to the injection molding direction and one perpendicular to it shown in Fig. 2a. Consequently, the specimens were labeled "Longitudinal" and "Transverse" as their length is parallel or perpendicular to the infusion direction.

The ASTM D7078 standard was also utilized for assessing the effect of the fabrication direction of plain weave textiles. To this end, two similar material, in terms of constituents' composition are considered. For confidentiality reasons, in the present work, these two materials are named GFRTP Material A and GFRTP Material B. The two materials consist of plain weave glass fibers with similar weight fraction (47% for the GFRTP Material A and 44% for the GFRTP Material B) placed in a polypropylene matrix. The two materials have a considerable unit cost difference; the GFRTP Material A is produced inside the EU while the GFRTP Material B in the Asian market. The specimens were cut according to the two main directions, one perpendicular to the warp and one perpendicular to the weft as shown in Fig. 2b.

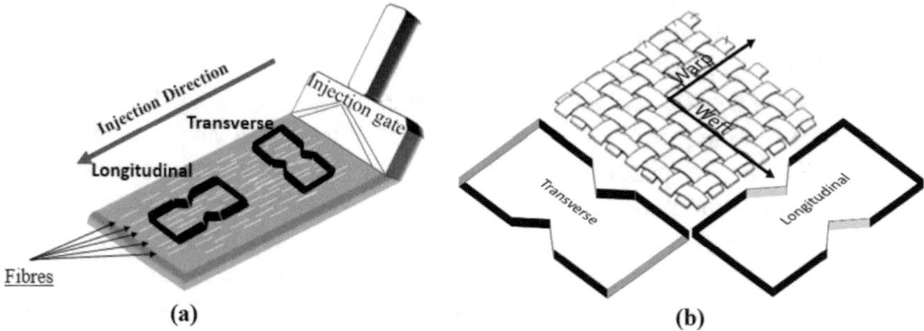

Fig. 2 Schematic representation of the correspondence of the Longitudinal and Transverse specimens with the fabrication direction

Table 2 Test matrix of the specimens of the mechanical tests conducted

Material	Description	Specimen designation	Testing speed (mm/s)	Replication tests
PP-GF-30	Short glass fiber reinforced polypropylene	Longitudinal	0.1	5
		Transverse		5
GFRTP Material A	Plain weave glass fiber reinforced polypropylene	Longitudinal (Warp)	0.1	5
		Transverse (Weft)		5
GFRTP Material B	Plain weave glass fiber reinforced polypropylene	Longitudinal (Warp)	0.1	5
		Transverse (Weft)		5

All the specimens were painted with a speckled varnish in order to create the proper pattern for strain inspection using the Digital Image Correlation technique (DIC). The complete test matrix is presented in Table 2.

3.2 Mechanical Testing and Data Acquisition

As seen above, a total number of 30 mechanical tests were conducted. An MTS servo electrical universal testing machine with maximum load capacity of 50 kN was utilized. The data acquisition rate was calibrated to 100 Hz. Alongside with the mechanical testing device, a Nikon DS5200 photo camera was positioned near the specimen in a way to capture 1 photo per second, allowing this way the DIC data acquisition for further analysis of the local and global strain field. One of the specimens mounted on the apparatus can be seen in Fig. 3. The captured images were elaborated with the Image J [27] software to increase the contrast and analyzed using the MatLab Ncorr [28].

The choice of the DIC analysis was not only performed to evaluate the strain distribution. According to previous research [23], even if the strain gauges are proposed by the ASTM D7078 standard [21] they fail to record the strain after a certain point of the execution of this particular testing method. The reason is the pine cracks which form at the section between the notches causing the de-cohesion of them.

As seen in Fig. 4, the shear strain distribution between the notches of the specimen is quite uniform throughout the execution of the mechanical tests. For calculating the shear strain, the region between the notches was isolated after the DIC analysis and elaboration and the γ_{xy} strain was obtained. For obtaining the stress values, the corrected values of force obtained are divided by the cross-sectional region between the notches tips.

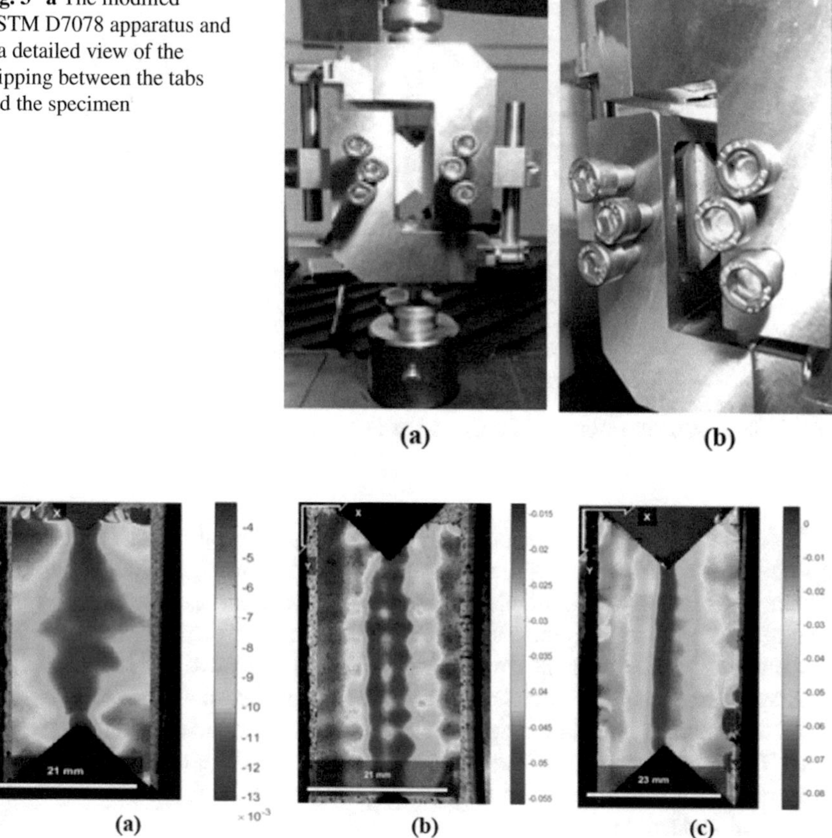

Fig. 3 **a** The modified ASTM D7078 apparatus and **b** a detailed view of the gripping between the tabs and the specimen

Fig. 4 Typical γ_{xy} shear strain distribution of the **a** PP-GF-30, **b** GFRTP Material A and **c** GFRTP Material B respectively

In addition to the above, for assessing the adequacy and the effectiveness of the guiding mechanism, 3 more tests were conducted without the use of the guiding-alignment system using specimens made of the PP-GF-30 material of the Transverse category. The choice was made in a way to investigate the less reinforced material's response. After the tests, the shear strain distribution and the uniformity of the shear deformation on the region between the notches was measured. In Fig. 5, a comparison between the shear strain distribution, moments before the final failure of the specimens tested with the modified apparatus (a) and the standard apparatus (b) is made. As seen there, the strain distribution is much more uniform, symmetrical and concentrated in the zone between the V-Notches (color dark blue) when the guiding cylinders are utilized. On the other hand, the shear strains out of the region between the notches are more intense in the case without the use of guiding cylinders, leading to the conclusion that the shear strain is less concentrated in the desired zone. This fact which signifies the importance of their implementation.

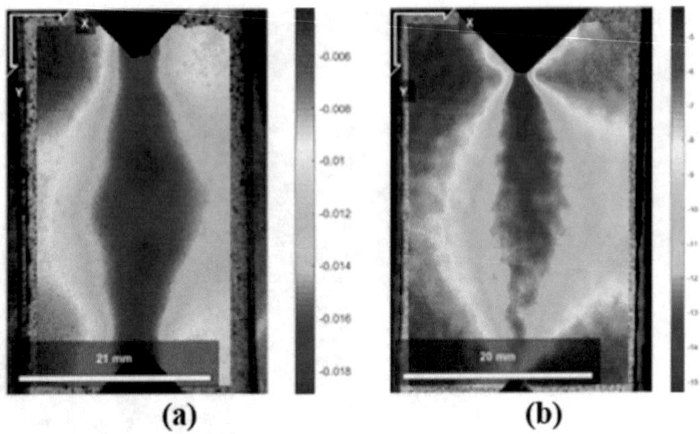

Fig. 5 Shear strain distribution of PP-GF-30 specimens with guiding cylinders (**a**) and without them (**b**)

Moreover, in order to understand more profoundly the significance of the guiding cylinders to the uniformity of the shear strain in the measuring region (between the notches), the region was subdivided into smaller regions aiming to assess the progression of the shear strain as the testing machine's crosshead displacement increased. The output of the assessment, even though delivered indicative results, pointed that the shear strain was almost the same for every region while a deviation was observed when the crosshead displacement exceeds 1 mm for the standardized testing apparatus. This phenomenon is resulted presumably by the application of stresses related to bending moments introduced to specimen that contribute to the development of stresses not related to shear.

4 Results and Discussion

4.1 Short Fiber Composites

The PP-GF-30 material proved to be quite fragile as its mechanical behavior was found to be non-linear elastic as it is presumably dominated by the polypropylene. The typical failure mode observed in both Longitudinal and Transverse specimens is an angled crack starting from the zone near the upper or lower notch and propagating almost instantaneously towards the direction of 45° of the loading direction. In Fig. 6a, a typical load-displacement curve obtained by one of the Transverse specimens along with photos recorded at various steps of the specimen deformation are presented. Consequently, after the elaboration of the obtained frames of the photo-camera, a comparison between the Longitudinal and Transverse specimens is

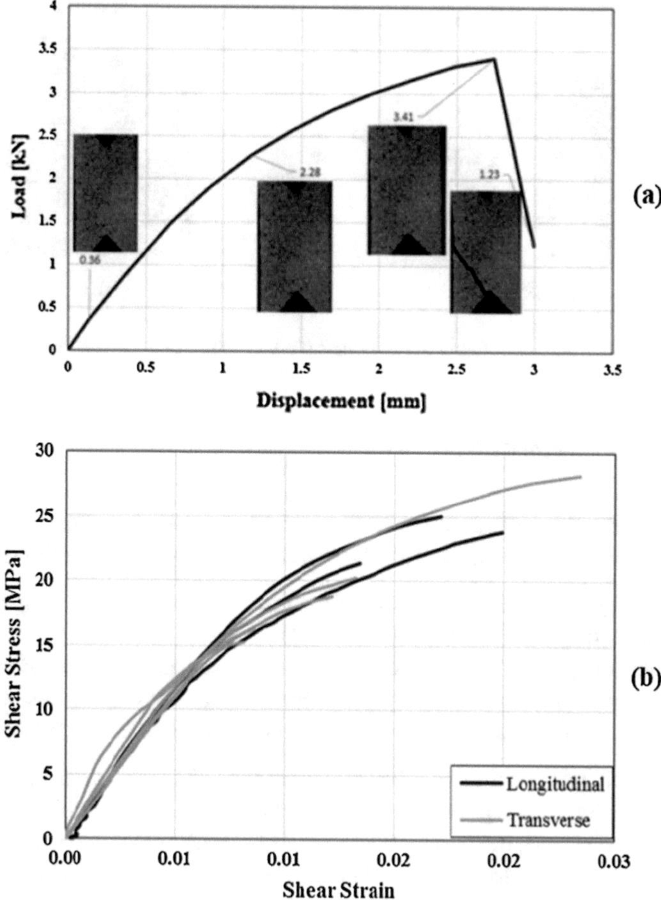

Fig. 6 **a** Typical load-displacement curve of a PP-GF-30 specimen tested according to the ASTM D7078 standard and **b** comparison of the in-plane shear behavior of Longitudinal and Transverse PP-GF-30 specimens

presented in Fig. 6b, while the averaged values of the in-plane shear strength and modulus may be seen in Fig. 7.

A first observation in both cases is that generally the longitudinal specimens tend to exhibit higher values of both strength and modulus, presumably due to the fiber alignment imposed by the manufacturing process as described in the previous sections of the present work. Nevertheless, observed is a significant standard deviation on the results leading to the conclusion that percentage of this fiber alignment is not the same in all of the longitudinal and transverse specimens, thus there must be a deviation on the percentage of the fibers aligned to the longitudinal or to the transverse direction of the injection of the plate from which the specimens were cut.

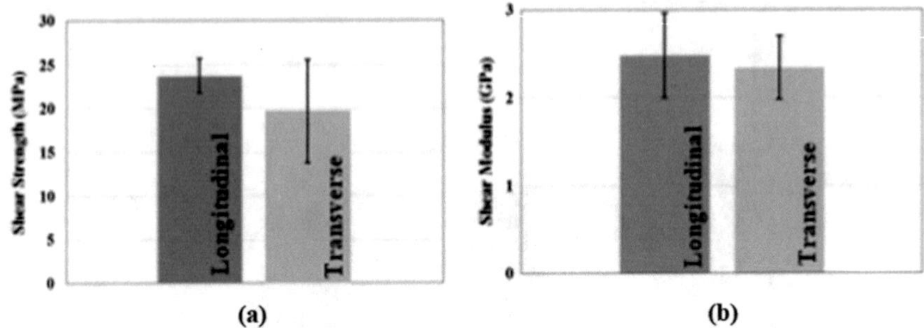

Fig. 7 Comparison of the average shear strength (**a**) and the in-plane shear modulus (**b**) of the PP-GF-30 material

4.2 Textile Composites

A representative load-displacement curve of each textile composite material is depicted in Fig. 8. This behavior may be divided into two parts, as follows;

- one which goes from the beginning of the test until the first load drop where the specimen is subjected to pure shear;
- right after this point, the slope of the curve increases due to the progressive fiber alignment which introduces other phenomena apart from shear such as flexural loading, friction between fiber strands, local compression and/or tension.

Starting with the first material (GFRTP Material A), the in-plane shear mechanical behavior is depicted in Fig. 8a while the corresponding behavior of the second in Fig. 8b. As can be seen, the two materials behave mechanically in a similar manner, exhibiting both the two sections of the curve described previously. Considering the above, the nature of the textile composites is the main reason of this particular

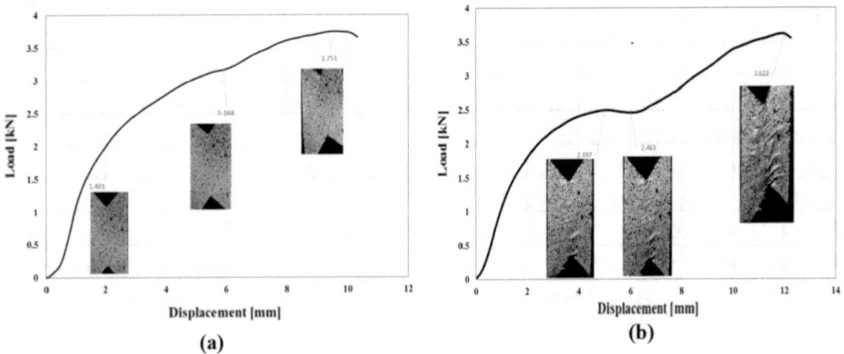

Fig. 8 Typical load-displacement curves of the **a** GFRTP Material A and **b** GFRTP Material B, correlated with frames of the deformed central section of the specimen

mechanical behavior and therefore the failure point of the curve should be carefully considered and observed in detail. In the present work, this point has been recognized when the first cracks are observed at the middle section of the specimen (between the V-Notches) and cause the load to drop slightly. By comparing the shear stress-strain curves of the two materials, it can be seen that for the second material (GFRTP Material A) such point is more easily distinguished.

In Fig. 9 the deformed specimens of the two materials after the execution of the shear tests are presented. It may be seen that, after the second part of the curve a combination of phenomena are observed in the region between the notches, that are highly related with the relative deformation of the fiber strands, leading to the development of pine leaf-shaped cracks before the initiation of the second part of the curve. Then, a combination of phenomena such as local buckling, fiber-matrix decohesion and fiber-fiber friction and several matrix cracks are influencing the mechanical behavior of the material.

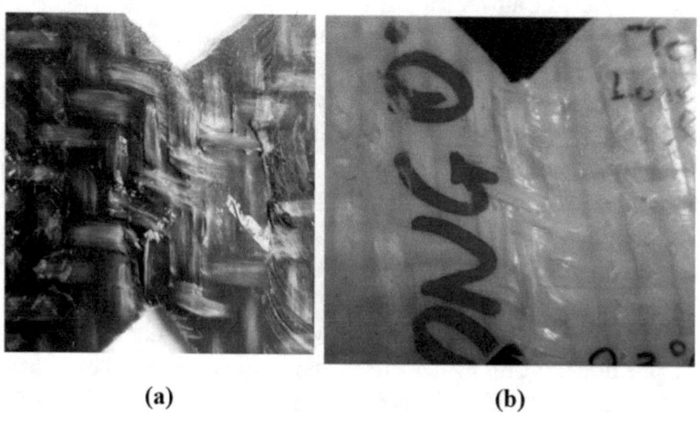

Fig. 9 Deformed surfaces of the GFRTP Material **a** A and **b** B, respectively

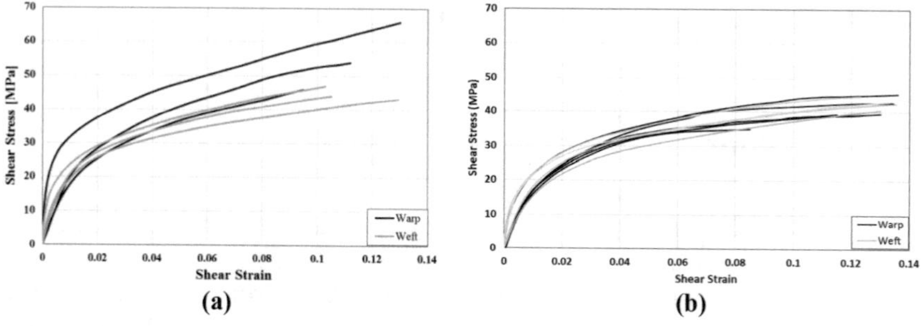

Fig. 10 Shear stress-strain curves of **a** Material A and **b** Material B, for specimens oriented along warp (Longitudinal) and weft (Transverse) directions, respectively

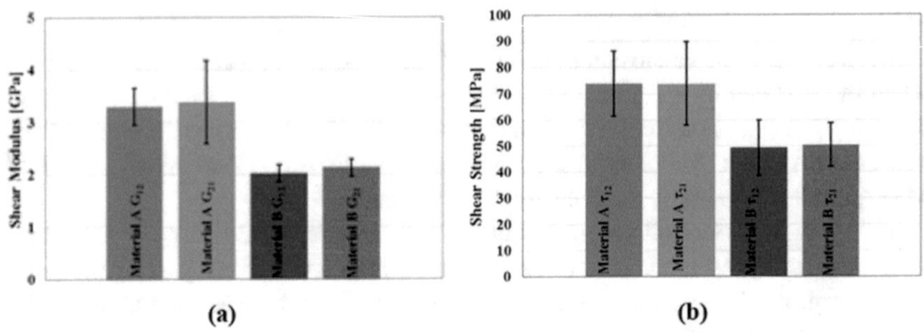

Fig. 11 Results for the two materials regarding **a** the shear modulus and **b** the shear strength

Considering the above, the stress-strain curves of the two materials, may be seen in Fig. 10a, b respectively, where only the first part of the trends shown in Fig. 8a, b respectively is depicted (pure shear part).

From the stress-strain curves it is obvious a certain difference between the 1–2 plane (black curves, warp) and 2–1 (grey curves, weft) plane properties of the Material A which are always accompanied with a certain dispersion. It is also noted that this difference is not so intense in the Material B. Nevertheless, even if the two materials (A and B) are having almost the same material base (matrix and fibers) with a slight difference on the fiber content, their shear behavior appears to be quite different. An overview of the mechanical properties of the two materials is presented in Fig. 11 where the properties of GFRP Material A are colored blue and those of the GFRP Material B are colored orange. As can be seen there, the GFRTP Material A demonstrates an augmented shear strength and modulus compared to the GFRTP Material B, presumably due to the difference between their fiber content (47 vs. 43%). Nevertheless, their difference between their strength is roughly 30% while the in-plane shear modulus of the Material B is almost the half of the Material A. These results are also explained by the observed variations of the strands of the glass fibers of the Material B as well as the not proper fiber alignment (Fig. 8b), a fact which indicates the influence of fabrication process errors.

5 Conclusions

In the present work, an improvement of the apparatus of the most promising and precise, in terms of accuracy, shear test for composite materials was presented and implemented, to authors' knowledge, for the first time on short fiber reinforced thermoplastics. In addition, this attempt appears to be among the few for the assessment of the shear behavior of plain weave glass reinforced polypropylene. The modifications made seem to contribute significantly on the specimen proper alignment during the test with respect to the loading direction and the strain field observed via DIC is very uniform.

Even if this test was not implemented before on short fiber reinforced composites, the results obtained appear to be highly repeatable and accurate in terms of both mechanical behavior and properties. Consequently, this testing method throughout this work may be considered as a benchmark on the characterization of the shear properties and mechanical behavior of these materials.

From the modified ASTM D7078 tests conducted for the textile glass fiber polypropylene (Materials A and B) the apparatus combined with the DIC analysis may provide the user with accurate and repeatable results. The results obtained applying the proposed apparatus enabled to assess the small differences in the behavior of the composites due to the alignment of the fibres caused by the matrix flow during injection or to a small difference in fiber content in the case of textile composites.

Finally, this work may be considered as a benchmark of a further use of this testing method for assessing the, so-easily-influenced by the fabrication, shear properties of not only UD composites but also chopped-fiber or textiles in a range of testing conditions. The presented apparatus may be considered as the first step towards further modifications that can contribute to a universal shear testing apparatus, fabricated with less removable parts, stiffer and more stable while being used in a range of temperatures and humidity conditions.

Acknowledgements This work has received funding by the national research project C.R.AB (Composites Research Abruzzo) under the auspicies of the Region of Abruzzo (Italy) and the European Union (PRO FESR Abruzzo 2014–2020—ASSE I-Attività I.1.1 e I.1.4, contract CAR n.2617-COR n.216522).

Authors gratefully thank Dipl.-Ing. Luca Glauco Di Genova for his contribution in the design of the testing fixture and the assistance during the tests as well as the Responsible of the machine shop of the Department, Mr. Cesare Michetti for the realization of the developed equipment.

References

1. Ishikawa T, Amaoka K, Masubuchi Y, Yamamoto T, Yamanaka A, Arai M, Takahashi J (2018) Overview of automotive structural composites technology developments in Japan. Compos Sci Technol 155:221–246
2. Ishikawa T (2014) Overview of CFRP (carbon fiber reinforced plastics) application to future automobiles. J Soc Autom Eng Jpn 68:4–11
3. Nayak NV (2014) Composite materials in aerospace applications. Int J Sci Res Publ 4–9
4. Fedulov BN, Antonov FK, Safonov AA, Ushakov AE, Lomov SV (2015) Influence of fibre misalignment and voids on composite laminate strength. J Compos Mater 49(23):2887–2896
5. Stamopoulos AG, Tserpes KI, Prucha P, Vavrik D (2016) Evaluation of porosity effects on the mechanical properties of carbon fiber-reinforced plastic unidirectional laminates by X-ray computed tomography and mechanical testing. J Compos Mater 50(15):2087–2098
6. Mangino E, Carruthers J, Pittaresi G (2007) The future use of structural composite materials in the automotive industry. Int J Veh Des 44:3–4
7. Hovorun TP, Berladir KV, Pererva VI, Rudenko SG, Martynov AI (2017) Modern materials for automotive industry. J Eng Sci 4(2):1–11
8. Chehroudi B (2015) Composite materials and their uses in cars part II: applications. PhD thesis

9. Launay A, Marco Y, Maitournam MH, Raoult I (2011) Constitutive behavior of injection molded short glass fiber reinforced thermoplastics: a phenomenological approach. Procedia Eng 10:2003–2008
10. Wang J, Nguyen BN, Mathur R, Sharma B, Sangid MD, Costa F, Jin X, Tucker CL, Fifield LS (2015) Fiber orientation in injection molded long fiber reinforced thermoplastic composites. In: Proceedings of the technical conference & exhibition, Orlando. Society of plastics engineers & society of plastics engineers ANTEC 2015, Florida, 23–25 Mar 2015
11. Gupta M, Wang KK (1993) Fiber orientation and mechanical properties of short-fiber-reinforced injection-molded composites: simulated and experimental results. Polym Compos 14(5):367–382
12. Zainudin ES, Sapuan SM, Sulaiman S, Ahmad MMHM (2002) Fiber orientation in short fiber reinforced injection molded thermoplastic composites: a review. J Inject Mold Technol 6(1):1–11
13. Chen F, Jones FR (1995) Injection molding of glass fiber reinforced phenolic composites: 1. Study of critical fiber length and the interfacial shear strength. Plast Rubber Compos Process Appl 23(4):241–258
14. Dowd FO, Levesque M, Gilchrist MD (2006) Analysis of fibre orientation effect on injection moulded components. Procedia IMechE Part B J Eng Manuf 20
15. Demircan O, Ashibe S, Kosui T, Nakai A (2004) Mechanical properties of biaxial weft-knitted and cross-ply thermoplastic composites. J Thermoplast Compos Mater 28:1058–1074
16. https://www.compositesworld.com/articles/a-comparison-of-shear-test-methods
17. Adams DO, Moriarty JM, Gallegos AM, Adams DF (2003) Development and evaluation of the V-notched rail shear test for composite laminates. U.S. Department of Transportation, Federal Aviation Administration, Office of Aviation Research, Washington
18. Adams DO, Moriarty JM, Gallegos AM, Adams DF (2007) The V-notched rail shear test. J Compos Mater 41(3):281–297
19. ASTM D5379/D 5379M – 98. Standard test method for shear properties of composite materials by the V-notched beam method. American Society of Testing and Materials
20. ASTM D 4255/D 4255M - 01. Standard test method for in-plane shear properties of polymer matrix composite materials by the rail shear method. American Society of Testing and Materials
21. ASTM D 7078/D 7078M - 12. Standard test method for shear properties of composite materials by V-notched rail shear method. American Society of Testing and Materials
22. Almeida Jr JHS, Angrizzani CC, Botelho EC, Amico SC (2015) Effect of fiber orientation on the shear behaviour of glass fiber/epoxy composites. Mater Des 65:789–795
23. Stamopoulos A (2017) A numerical methodology for predicting the mechanical properties of unidirectional composite laminates with pores: evaluation of porous CFRP specimens using X-ray CT data and an artificial neural network. PhD thesis, University of Patras, Mechanical Engineering and Aeronautics Department, Laboratory of Technology and Strength of Materials, Greece
24. Taheri-Behrooz F, Moghaddam HS (2018) Nonlinear numerical analysis of the V-notched rail shear test specimen. Polym Test 65:44–53
25. Gude M, Hufenbach W, Andrich M, Mertel A, Scirner R (2015) Modified V-notched rail shear test fixture for shear characterisation of textile-reinforced composite materials. Polym Test 43:147–153
26. https://www.skf.com/binary/68-245746/Linear-bearings-and-units---4182_2-EN(1).pdf
27. Ferreira T, Rasband W (2010) ImageJ user guide. IJ1.46r. Html version
28. Blader J, Antoniou A (2014) Ncorr instruction manual. Version 1.1. Georgia Institute of Technology

An Approximate Approach for the Verification of Process Plans with an Application to Reconfigurable Pallets

Massimo Manzini and Marcello Urgo

Abstract The manufacturing sector has to be able to manage high-variety and low-volumes per product, causing the adoption of a dedicated production system/cell to be unfeasible. In this context, reconfigurable pallets and flexible fixtures are enablers to manage product variety and volume variability. Namely, as a pallet is reconfigured, the associated part program needs to be verified to check for possible collisions between the tools and the new machining environment. An approach is proposed to verify the machinability of a pallet configuration given an existing part program. The approach grounds on an approximated collision check method exploiting a 3D representation of the machining environment (fixtures and parts). The approach is validated through an application to a realistic use case and the comparison with the results of a traditional collision check approach.

Keywords Process planning · Process verification · CNC

1 Introduction

High-variety and low-volume are typical characteristics of industrial production today, driven by the proliferation of models or variants for the products. In this context, companies have to cope with these environmental constraints through systems able to cope with a high variety of parts and frequent set-ups, with the aim at guaranteeing a reasonable utilization for the equipment. This approach is the only viable one, in comparison to the implementation of production systems dedicated for each specific product, whose feasibility is endangered by the very high variety

M. Manzini (✉) · M. Urgo
Manufacturing and Production Systems Laboratory, Mechanical Engineering Department, Politecnico di Milano, Via La Masa 1, 20156 Milan, Italy
e-mail: massimo.manzini@polimi.it

M. Urgo
e-mail: marcello.urgo@polimi.it

E. Ceretti and T. Tolio (eds.), *Selected Topics in Manufacturing*, Lecture Notes in Mechanical Engineering,
https://doi.org/10.1007/978-3-030-57729-2_8

of parts. Besides being a typical environmental condition for the production of new products, the described situation also affects production activities devoted to different phases of the life-cycle of a product. An example comes from the automotive industry where, as a model of a car exits from the series production phase, dedicated production systems or cells for its components are not likely to be feasible to maintain. At the same time, OEMs have to guarantee the supplying of spare parts for a considerable time (e.g., 10 years), thus, they typically outsource the production of spare parts to companies facing environmental conditions very similar to the ones described above. Another example comes from the application to the remanufacturing, where companies providing refurbished products have to be able to work/rework parts using equipment and processes that could be different from the original ones. Also in this case, although the volumes are not necessarily low, the variety of the product mix is very high since rework activities could request shorter processing times. As a consequence, production systems able to cope with a wide range of parts are a clear requirement.

Flexible and/or reconfigurable production systems are the main paradigms to cope with these requests and guarantee a reasonable utilization factor for a multi-purpose system [15, 19] according to the *co-evolution* principle [16], i.e., the need of modifying the configuration of a production system or its components together with the changes affecting the products or the processes. Reconfigurability and flexibility have been mostly addressed in the design of manufacturing systems in general, with a special focus on machine tools. Nevertheless, these two paradigms are also meaningful and relevant with respect to fixtures (e.g., referring to machining and/or assembling of parts) representing one of the most relevant enablers to manage high-variety of products. A possible solution to this requirement is provided by zero-point clamping system [17], where specifically designed devices allow the fast and reliable reconfiguration of fixtures. Preconfigured baseplates hosting different sets of fixtures can be rapidly mounted and unmounted onto a standard pallet tombstone, guaranteeing the referencing of the parts without the need to align and check the modular fixtures again. Thus, enabling a fast and reliable pallet reconfiguration.

As the pallet configuration changes, a new part program has to be devised or, at least, the reuse of the previous part programs has to be verified to check the machinability of the new pallet configuration and avoid collisions between the tool and the fixtures. This verification could be extremely time-consuming and constitutes one of the main limitations to the adoption of fast pallet reconfiguration technologies in the industry.

In this paper, we propose an approach pointing in this direction. The approach is able to perform an approximate verification of the machinability of a part considering the environmental conditions and constraints, i.e., the fixtures hosting the part, the presence of other pieces of fixture in the working environment, the characteristics of the machining tool. The approach grounds on a collision detection method approximating the volume the tool will sweep while executing the part program and checks possible collisions with other elements in the working environments (i.e., fixtures, other parts, elements of the machine).

The advantage of the proposed approach is to accept approximation in change of speed, thus constituting a preliminary analysis step being able to check a high variety of alternative fixture configuration, while relying on traditional approaches based on the simulation of the machining process for the final validation of the part program. In this paper we provide a feasibility analysis of the proposed approach in a realistic case and a comparison with more traditional solution approaches.

Outline The paper is organized as follows: Sect. 2 provides an analysis of the literature, also highlighting the main advancements, while the complete problem statement is presented in Sect. 3, where the process plan verification problem is formalized. In Sect. 4, the solution approach is described through its three steps. The viability of the approach is demonstrated through the application to an industrial problem in Sect. 5. Conclusions and future development directions are provided in Sect. 6.

2 State of Art

In the last years, great importance has been given to the development of technological solution aiming at facing environmental constraints as the high-variety and low-volume situation [3, 14]. These solutions always ground on reconfigurability and flexibility concepts, as the zero-point clamping system one. The main limitation in the adoption of this technology in pallet configurations lies on the verification of the machinability of the new pallet using the part program used for the previous pallet.

The machinability can be verified using simulation tools, e.g., Vericut (https://www.cgtech.it/products/about-vericut/) or Moduleworks (https://www.moduleworks.com/) able to virtually reproduce the material removal process but also to check the presence of collisions between cutting tool and other elements in the working environment (e.g., pallet structure and fixtures mounted on it). In doing this, they consider the volumes occupied by the different elements in the working environment and divide them into sub-elements, then, they check whether these sub-elements collide or not. Another possibility is to verify the machinability with a collision detection approach [12] able to analyze in detail the movement of the cutting tool with respect of the working environment. Different approaches have been presented: the ones that study the surface's property of the objects involved [2], the ones based on the relative distance between objects [20], and the ones that simplify the shape of the objects involved [4, 8, 13, 18].

All these methods are complex and time consuming due to the need to represent and handle the entire volume of the elements, and the level of detail provided. A rapid and efficient pallet inspection could increase the system adaptability to market requests by boosting the adoption of reconfigurable pallet configurations. For this reason, the aim of this paper is to propose an approximate approach to be used in a preliminary analysis.

In this sense, a valuable approach is to verify the machinability by checking the accessibility of the cutting tool to the part mounted on the pallet [1, 9, 10]. In this case, the main drawback is that the verification of accessibility is not focused on the volume occupied by the elements in the working environment and, thus, it does not guarantee the absence of collisions during the process.

In this paper, we present a fast approach able to evaluate the machinability of a new pallet using a given part program by approximating the elements inside the working environment with their convex hull and then evaluating their overlapping. In this way, the approach considers the volume of every element by evaluating the surface surrounding it without the complexity and the computational effort required by the previous approaches.

3 Problem Statement

Designing a pallet configuration usually follows the steps shown in Fig. 1. Firstly, a set of configurations are defined in terms of the set-ups of the parts, their position in the working environment, the position of the associated fixtures as well as additional pieces of fixture (e.g., plates or columns) to be assembled onto the pallet. For each of these configurations, a part program has to be defined and checked. For this check, a simulation tool is typically used (https://www.cgtech.it/products/about-vericut/) (https://www.moduleworks.com/), providing an environment where, after the definition of the configuration, the part program and the characteristics of the machines and the CNC, the trajectory of the tool is simulated and a check for collision is operated. Although effective, this simulation step is usually time consuming

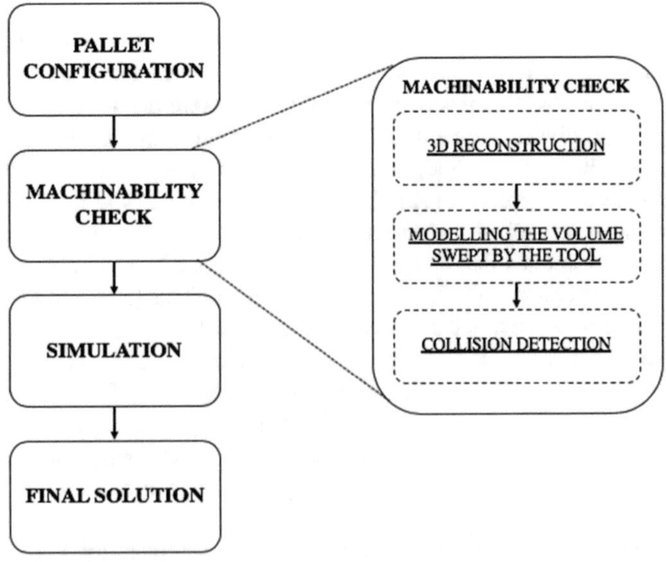

Fig. 1 Workflow of the approach

and it often requires a wide range of detailed information that, in many of the conditions previously described, could not be available (e.g., the specific machine tool to be used could be unknown). To overcome this limitation, the proposed approach is aimed at being an alternative for this simulation step, providing an approximate verification for the machinability through a collision detection approach.

Traditional process simulation and verification approaches follow a simulation strategy, calculating the relative positions of the tools and the parts to machine in the machine environment. For each point in this discretization, a collision check is operated between the 3D model of the tool and the 3D model of the working environment (i.e., parts, fixtures, elements constituting the machine, etc.).

In this scheme, a very high number of collision checks have to be done and this obviously has a significant impact on the computation time. To mitigate this impact, we introduce a preliminary evaluation step (see Fig. 1) where a limited number of sampled points is considered. Starting from these, the 3D representation of the tool (and the attached mandrel) and its sequential positions in the machine space are considered to calculate the convex hull of these volumes. This convex hull is an approximation of the space swept by the tool while executing the part program. As the number of positions considered increases, the convex hull tends to match exactly the real volume swept by the tool.

More formally, the volume V_{Btot} represents the space occupied by the elements inside the working cube (e.g., fixtures, structural components of the machine tool, parts not to be machined) calculated as the union of the volume occupied by each of these elements, $V_{Btot} = \cup_{b\in\{1,\ldots,B\}} V_b$. V_P is the volume occupied by the parts to be machined; it is considered separately from V_{Btot} since a collision with it is expected. In addition, the volume V_C identifies the convex hull of the volumes occupied by the cutting tool and the mandrel in a set of points during the machining process. Grounding on this, it is possible to approximately verify the existence of collisions between the tool and the fixtures by checking for collisions between V_{Btot} and V_C. Nevertheless, as stated before, V_C provides an approximation of the real swept volume and, thus, the absence of intersections does not guarantee for the absence of collisions.

4 Solution Approach

The approximate approach follows a three-phase implementation. In the first phase, the volumes occupied by the part types to be worked, the pallet, the fixtures blocking the parts and the cutting tool are derived (Sect. 4.1). Then, the proposed approach for the definition of the volume swept by the tool during the process is presented (Sect. 4.2). Finally, (Sect. 4.3) the *collision detection* approach is described to verify the machinability by evaluating the overlapping between the considered 3D objects.

4.1 3D Space Reconstruction

The volumes V_b, $\forall b \in \{1, \ldots, B\}$ occupied by the elements blocked inside the machine (e.g., parts not to be machined, the pallet and the associated fixtures blocking the parts) are derived directly from their 3D representations and positioned in the working environment.

Given the 3D representation of an object, its position inside the working cube of the CNC machine is defined through a tree of origins, i.e., a sequence of coordinate system origins and their relative positions.

The first origin considered is the *origin of the machine coordinate system*, located at point (0, 0, 0) in the working cube. The *pallet origin* represents the position of the pallet inside the working cube as a reference to the *origin of the machine coordinate system*. Then, a set of fixtures are mounted on the pallet blocking the parts to be worked. The positions of fixtures and parts are defined with their *fixture origins* and *part origins* with reference to the *pallet origin* and the distance from it.

The distances and the translations between these origins are defined through a set of *Homogeneous Transformation Matrices* (*HTM*s). An *HTM* is a matrix $R_{4\times4}$ given by the product of three rotational matrices (one for each principal axes) and the translation matrix. In general, it is represented as

$$R_{4\times4} = \begin{bmatrix} D_{3\times3} & T_{3\times1} \\ P_{1\times3} & s \end{bmatrix} \tag{1}$$

where D is the rotation matrix, T is the translation matrix, P is the perspective matrix and s is a scale factor. The position of each component inside the working cube of the CNC machine is obtained as a sequence of *HTM*s applied to the *origin of the machine coordinate system*. More formally, the space occupied by element $b \in B$, e.g., the parts, the pallet and the fixture, is a function of its position p_b and its 3D representations, thus, $V_b = V_b(p_b, 3D_b)$, where the position p_b is identified through a sequence of *HTM*s. The union of these volumes represents the volume of the entire set elements, $V_{Btot} = \cup_{b\in\{1,\ldots,B\}} V_b$.

As a consequence, for the identification of the volume V_{Btot} occupied by the elements inside the working cube, the information about the characteristics of the machine, e.g., the size of the working cube, are not needed. The only information needed is the relative position of the elements and their 3D models. This makes the approach independent from the complete definition of the machine environment and suitable for an evaluation grounding on the pallet configuration and the part program only.

4.2 Modelling the Volume Swept by the Tool

During the execution of the process, the tools moves with respect to the pallet thus, in order to check for possible collisions between the tool and the fixtures, the volume swept by the cutting tool has to be computed.

The first step of the approach derives the sequential positions of the tool in the working cube grounding on the instructions in the part program. This phase is usually called *post processing* and needs a dedicated interpreter able to translate the part program into the movements of the elements of the machine, according to the specific CNC and machine architecture. Post processing tools have to cope with different G-Code languages and dialects with respect to the specific manufacturer (e.g., Fanuc, Siemens, etc.). When addressing the verification of the part program through simulation, software packages like Vericut (https://www.cgtech.it/products/about-vericut/) or Moduleworks (https://www.moduleworks.com/) provide their own post processor. Since many interpreters are available for post processing a part program and this is not the core part of this work, we assume to start from the output of a post processing phase, thus, from the sequence of the positions of the tool.

Starting from this information, a subset of positions is selected to reconstruct the volume swept by the tool. For each of these positions, the 3D model of the tool, together with the moving parts of the machine (e.g., the mandrel), is considered. The convex hull of these 3D models is used as a proxy of the volume swept by the tool. More formally, the volume associated to the tool is a function of its stereolithography (STL) representation and the set of positions selected from the trajectory, $V_C = V_C(PP, STL_C)$.

The number of positions sampled has a clear impact on the degree of approximation of the obtained convex hull. A first option consists in just considering the initial and final positions for each segment of the tool's trajectory, usually derived from a single instruction in the part program. This choice perfectly suits the approximation of linear movement of a CNC machine's axes, where the convex hull of the initial and final location of the tools just depends on these two positions. On the contrary, where non-linear trajectories and/or rotational axes movements are evaluated, simply considering the initial and final positions entails a very poor approximation. To overcome this issue, additional positions can be sampled between the initial and final ones.

Limiting the analysis to a traditional architecture of 4-axes CNC machine with horizontal mandrel, we can have two different cases: linear trajectories of the tool only relying on the linear axes; non-linear trajectories or movements involving the 4th rotational axes. Starting from the first one, the example in Fig. 2a represents the trajectory of a tool in the XY plane. It contains two segments, a linear one (AB) and a non-linear one (BC). The application of the two-points sampling is shown in Fig. 2b, where the convex hull is represented by the light green region. Clearly, the approximation of the non-linear segment of the trajectory is very poor, while it is acceptable for the linear one. By sampling 4 points from the segment AB and other 4 points from arc BC, it is possible to obtain a better approximation (Fig. 2c).

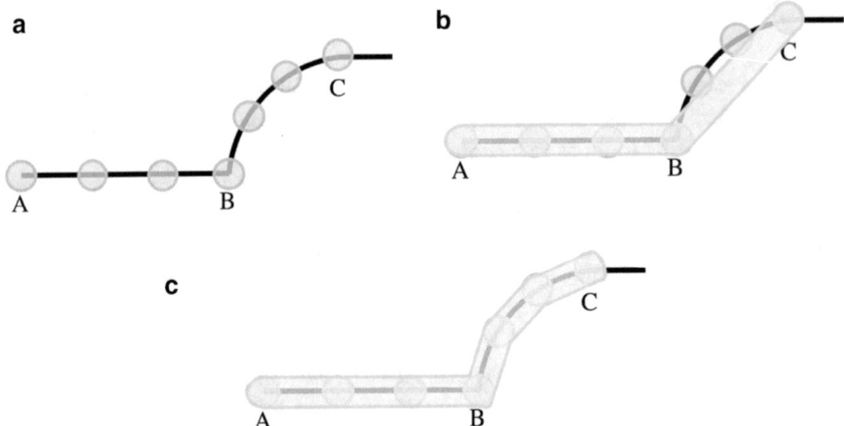

Fig. 2 **a** First row on the left, example of trajectory with different positions of the tool; **b** first row on the right, first sampling option; **c** second row, second sampling option

On the contrary, in the case when the fourth axis rotates, e.g., due to a variation of the tilt or the lead angle, we can consider this movement as a non-linear segment of the trajectory and apply a more frequent sampling as described.

Therefore, the number of sequential positions considered to compute the volume V_C increases, a better approximation of the real trajectory is obtained. On the other side, sampling many positions will require many STL elements to be considered in the calculation of the convex hull and, thus, a bigger computational effort.

The quality of the approximation is also affected by the sampling strategy. In the case of non-linear segments, the minimization of the chordal error, defined as the difference between the ideal arc section and the approximation using segments, is considered. In particular, given the number of positions to be sampled on a circular segment, their coordinates are chosen by following the Tustin interpolation method [11].

In the approach presented in this paper, two different sampling strategies for linear and non-linear segments of the tool's trajectory are considered. A first strategy only considers the initial and final position of the curved path, while the second one partitions the path into a given number of segments. For example, Fig. 2c, 4 positions have been chosen and 3 segments identified. Hence, the STL representations of the tool and the other moving parts of the machine are considered. For each segment, the associated convex hull is identified using the STL representations on its boundary points. Then, the volume swept by the tool during the process is approximated through the union of these convex hulls.

Sampling more than 2 positions from the non-linear segments of the trajectory helps to obtain a better approximation of the swept volume. An investigation of the impact on the performance of the approach varying the number of points sampled in the curved segments of the trajectory will be carried out in Sect. 5.

We implemented this first part of the approach using C++ code together with Matlab. In particular, a script in C++ has been developed to read the trajectory of the tool and identify the set of positions to be considered. The current version of the approach is limited to linear and circular/elliptical segments. Then, we use Matlab for managing the STL representations of both the elements in the working cube and the cutting tool. The convex hulls and their union have been implemented with MATLAB's function *boundary*, giving in output a convex hull as an STL representation.

4.3 Collision Detection

Once V_{Btot} and V_C have been defined, their possible intersections have to be checked. To this aim, we operate on surfaces rather than volumes considering S_{Btot} and S_C, the boundary surfaces for V_{Btot} and V_C respectively. This provides a more agile representation of the 3D objects and a faster calculation of their collision. Notice that, when referring to simulation approaches for machining operations (https://www.cgtech.it/products/about-vericut/) (https://www.moduleworks.com/), they usually adopt an approach explicitly considering the 3D volumes of the tool and parts. This is driven by the need of simulating the process and, consequently computing the portion of the part to be machined. In our case, since we do not need to simulate the process but simply check for collisions, considering just the boundary surfaces instead of the whole volumes is a viable approach.

Once the two boundary surfaces S_{Btot} and S_C have been identified, a collision check is operated by verifying whether they overlap or not. Through the described approach, the possible collisions between S_{Btot} and S_C are evaluated. If a collision is detected, then the selected pallet (fixture) configuration is not suitable for the execution of the machining process. On the contrary, if no collision is detected, since the approach is approximated, the machinability cannot be guaranteed and, hence, further analysis must be carried out, e.g., through a detailed simulation of the machining process.

We implemented the second part of the approach using the C++ language. In doing this, we take advantage of the library *V-Collide* (https://gamma.cs.unc.edu/V-COLLIDE/), exploiting a collision detection method for arbitrary polygonal objects. This library provides a function that takes as input the STL representation of a series of objects, their positions in the space and evaluates the possible collisions among them. In particular, we give in input to the library the list of STL representations and the convex hull identified.

5 Application Case

The approach presented in Sect. 4 has been tested on the production process of an automotive component (Fig. 3a) that has to undergo a machining process. We focus the analysis on one of the set-ups to machine the part and consider a possible pallet configuration, different from the one originally used, with nine parts hosted on a column with three surfaces equipped with fixtures, as shown in Fig. 3c. The details of the fixturing solution are depicted in Fig. 3b. The machining process requires different machining operations, i.e., milling, drilling and boring. Some of the surfaces and holes, as shown in Fig. 3a, are not perpendicular to the fixturing surface, thus, the drilling tool has to approach the pallet taking advantage of the 4th axis of the CNC machine.

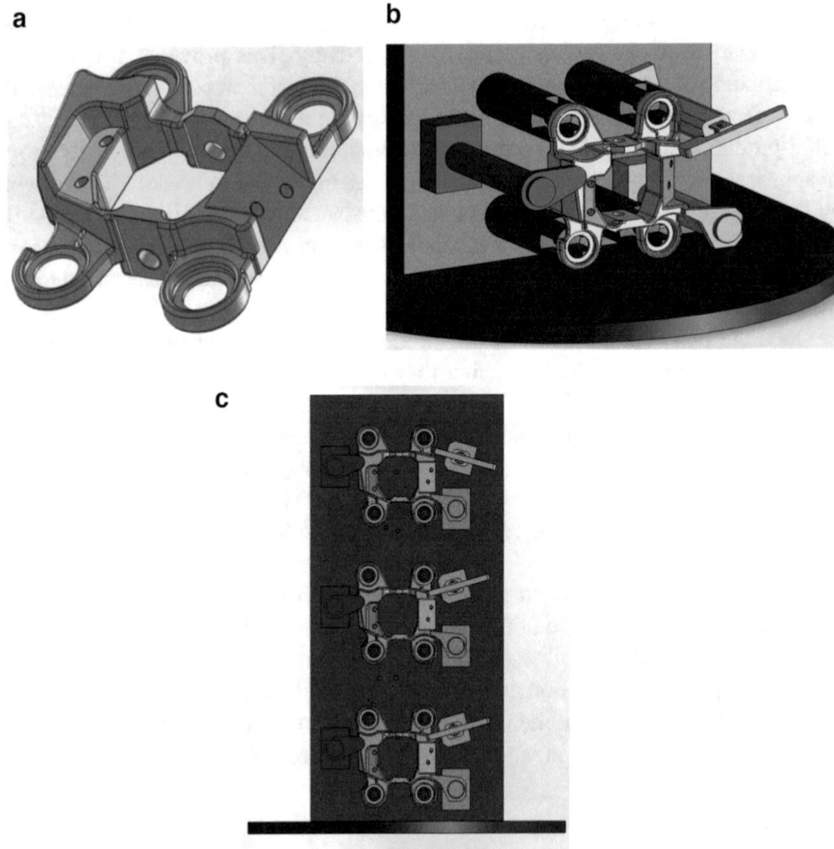

Fig. 3 **a** First row on the left, the automotive component; **b** first row on the right, details of the fixtures; **c** second row, the pallet configuration

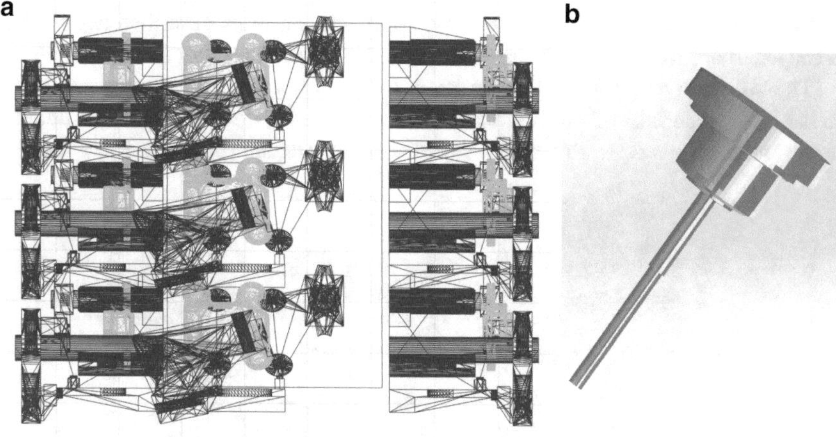

Fig. 4 **a** On the left, STL representation of parts and fixtures; **b** on the right, STL representation of the tool

Grounding on this, the described collision detection approach is used to check the machinability of the described pallet configuration with the available part program. To this aim, we firstly define volume V_{Btot} as the union of the volumes occupied by the pallet and the set of fixtures blocking the parts (Fig. 4a). In Fig. 4, the pallet, the set of fixtures (both in dark blue) and the set of parts (in light blue) are depicted using their STL representations.

Hence, the volume V_C swept by the tool is derived starting from the trajectory and the orientation obtained from the part program. For each of the considered operations, the specific STL representation of the tool used is considered (Fig. 4b). After the selection of the positions from the trajectory, the volume V_C is identified as the union of the convex hull of the positions of the tool in those points.

We consider two different sequences of operations, namely OP1 and OP2. In the first sequence, 4 end milling operations for each part in two faces of the pallet are executed, together with the rapid movement from a feature to another on the same part (3 for each part) and from a part to another (2 movements for each face). In the second sequence, we consider the execution of 4 end milling operations involving the translation of the tool according to the Y-axis of the machine tool and a rotation of the B-axis to move from a face of the pallet to another. For both the sequences we consider the application of the presented approach with two different sampling strategies, both minimizing the chordal error. The first strategy only considers the initial and final positions for both linear and curved segments of the trajectory, while the second strategy samples 4 positions in the curved segments only.

The proposed approach has been compared with a traditional sampling approach where a very small sampling interval is used, thus, obtaining a high number of points and, consequently, a very detailed description of the original trajectory. For each of these positions, the collision between the cutting tool and the other elements (pallet and fixtures) is operated, using the same approach and tools described above. This is the usual approach used for the simulation of a machining process.

The collision analysis for OP1 is given in Fig. 5a, where the volume associated to the tool executing the milling operation on two different surfaces of the pallet is represented in green. In this case, no collision is detected between V_{Btot} and V_C,

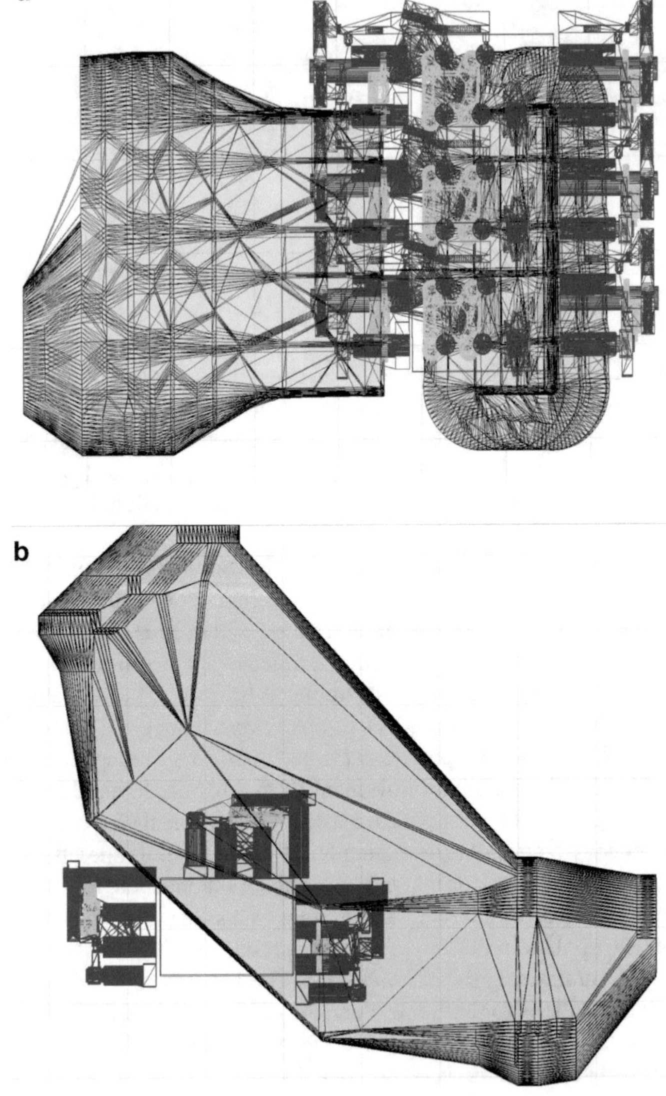

Fig. 5 **a** First row, sequence OP1; **b** second row, sequence OP2 from above

thus, the verification of the machinability of the pallet configuration has a positive outcome.

The collision analysis of OP2 is given in Fig. 5b (view from above) and Fig. 6a (lateral view). In this case, an overlap between the blue and red volumes is detected. A detail of the intersection between the two volumes is shown in Fig. 6b, with the overlapping highlighted in purple.

The results of the analysis are summarized in Tables 1 and 2. In particular, for the sequence OP1 (see Table 1), we sampled 200 and 400 positions with the first and the second strategy, respectively. Instead, for the sequence OP2 (see Table 2), we sampled 13 and 26 positions on the tool's trajectory. The V_C associated to OP1 counts 11,632 facets with the first strategy, and 15,632 facets with the second one. The OP2 is represented through 2124 and 2740 facets with the first and second

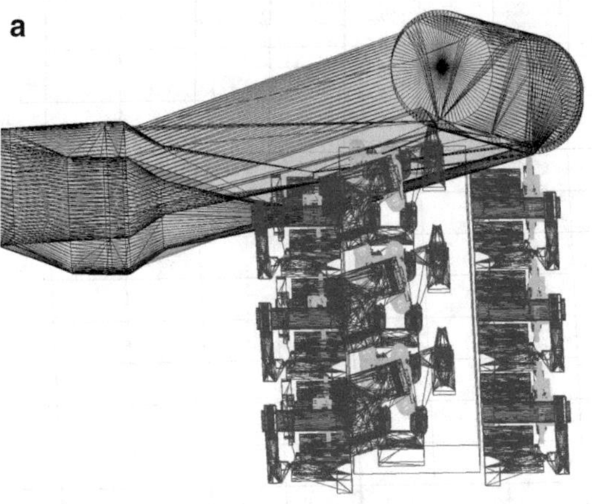

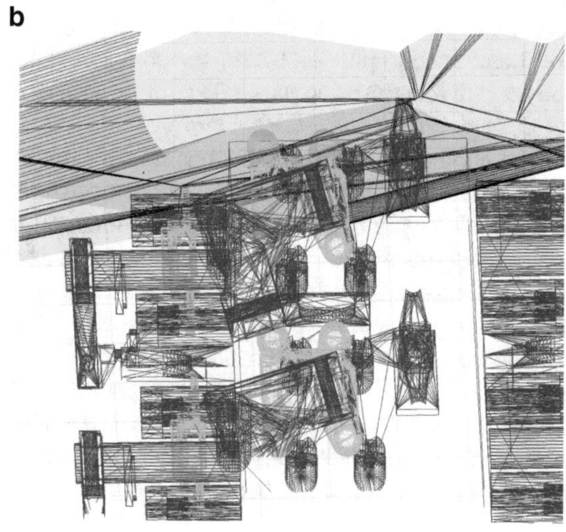

Fig. 6 **a** On the top, OP2, lateral view; **b** on the low, OP2, a detail of lateral view

Table 1 Results of the application case OP1

	OP1			
	Positions sampled	Triangles	Vertices in collision	Execution time (s)
Approximate approach: 1st strategy	200	268,534 + 11,632	0	32.13 + 2.89
Approximate approach: 2nd strategy	400	268,534 + 15,060	0	57.2 + 8.98
Continuous sampling	37,740	268,534 + 225,987,120	0	28.06

Table 2 Results of the application case OP2

	OP2			
	Positions sampled	Triangles	Vertices in collision	Execution time (s)
Approximate approach: 1st strategy	13	268,534 + 2124	258	1.95 + 1.41
Approximate approach: 2nd strategy	26	268,534 + 2740	385	3.02 + 3.30
Continuous sampling	2452	268,534 + 14,682,576	75,940	11.94

strategy, respectively. The V_{Btot} is the same for both strategies and both sequences, represented through 268,543 facets. The number of facets to be analyzed impacts on the execution time for the evaluation varying from 3.36 s (1.95 s for the convex hull identification plus 1.41 s for the collision checking) to 66.18 s (57.2 s for the convex hull identification plus 8.98 s for the collision checking) on a computer with 2.4 GHz processor and 8 GB memory.

The number of sampled positions impacts on the quality of the approximation, that can be represented with the number of vertices in collision. In particular, with the first strategy, a collision between the tool and the fixture has been detected in 258 vertices; instead, with the second strategy two collisions have been detected, the first one between the cutting tool and the fixture in 275 vertices, and the second one between the tool and the pallet in 110 vertices. It means that the collision detection approach takes advantage of the additional positions sampled according to the second strategy for the most critical region of the trajectory, the one in which the tool rotates.

The results of the traditional continuous sampling of the trajectory validates the ones of the approximated approach with the identification of the same collisions between the tool and the pallet, and between the tool and the fixture (see Tables 1 and 2). As expected, the associated computation times are different. In particular, considering only sequence OP2, the traditional approach spent 11.94 s for the collision evaluation, more than the time effort requested by the approximate approach for both strategies. Instead, considering only the sequence OP1, the time effort requested by the continuous sampling approach (28.06 s) is comparable with time spent by the approximated approach using the first strategy (35.02 s) and shorter than the second strategy one.

6 Conclusion

In this paper, we presented an approximate approach aiming at evaluating the machinability of a pallet configuration with a given part program. This approach is able to approximately check for possible collisions in the working environment, by reconstructing the volume swept by the cutting tool during different operations and evaluating its overlapping with the other elements in the machining environment.

The approach has been compared with a traditional simulation/verification approach for validation and to assess the benefits in of computational and time efforts. The results in this sense are optimistic when the number of positions sampled from the tool's trajectory are not many, thus, when the volume V_C is not too complex. Indeed, if we sample a high number of positions, the time effort requested for the identification of the convex hull is higher than the traditional approach one. Grounding on this, an investigation on the sampling strategies is needed to understand how the number of sampled positions impacts on the performance of the tool and what is the trade-off between the complexity of the convex hull and the approximation level of the collision evaluation.

Acknowledgements This research has been partially funded by Erasmus+ *Virtual Factory Learning Toolkit* Project, grant agreement no. 047094.

References

1. Borgia S, Pellegrinelli S, Petrò S, Tolio T (2014) Network part program approach based on the STEP-NC data structure for the machining of multiple fixture pallets. Int J Comput Integr Manuf 27(3):281–300
2. Ding XM, Fuh JY, Lee KS (2001) Interference detection for 3-axis mold machining. Comput Aided Des 33(8):561–569
3. ElMaraghy H, Wiendahl H-P (2009) Changeability—an introduction. In: Changeable and reconfigurable manufacturing systems, pp 3–24
4. Ho S, Sarma S, Adachi Y (2001) Real-time interference analysis between a tool and an environment. Comput Aided Des 33(13):935–947
5. https://www.cgtech.it/products/about-vericut/
6. https://gamma.cs.unc.edu/V-COLLIDE/

7. https://www.moduleworks.com/
8. Lee YS, Chang TC (1995) 2-phase approach to global tool interference avoidance in 5-axis machining. Comput Aided Des 27(10):715–729
9. Pellegrinelli S, Cenati C, Cevasco L, Giannini F, Lupinetti K, Monti M, Parazzoli D (2015) Design and inspection of multi-fixturing pallets for mixed part types. Procedia CIRP 36:159–164
10. Pellegrinelli S, Orlandini A, Pedrocchi N, Umbrico A, Tolio T (2017) Motion planning and scheduling for human and industrial-robot collaboration. CIRP Ann Manuf Technol 66:1–4
11. Suh S-H, Kang S-K, Chung D-H, Stroud I (2008) Theory and design of CNC systems. Springer
12. Tang TD (2014) Algorithms for collision detection and avoidance for five-axis NC machining: a state of the art review. Comput Aided Des 51:1–17
13. Tang TD, Bohez EL, Koomsap P (2007) The sweep plane algorithm for global collision detection with workpiece geometry update for five-axis NC machining. Comput Aided Des 39(11):1012–1024
14. Terkaj W, Tolio T, Valente A (2009) Designing manufacturing flexibility in dynamic production contexts. In: Design of flexible production systems. Springer, pp 1–18
15. Tolio T (2009) Design of flexible production systems. Springer
16. Tolio T, Ceglarek D, ElMaraghy HA, Fischer A, Hu SJ, Laperriere L, Newman ST, Váncza J (2010) Species—co-evolution of products, processes and production systems. CIRP Ann Manuf Technol 59(2):672–694
17. Urgo M, Terkaj W, Cenati C, Giannini F, Monti M, Pellegrinelli S (2016) Zero-point fixture systems as a reconfiguration enabler in flexible manufacturing systems. Comput Aided Des Appl 13(5):684–692
18. Wang QH, Li JR, Zhou RR (2006) Graphics—assisted approach to rapid collision detection for multi-axis machining. Int J Adv Manuf Technol 30(9):853–863
19. Wiendahl H-P, Hernández R (2001) The transformable factory—strategies, methods and examples. In: 1st international conference on agile, reconfigurable manufacturing, Ann Arbor
20. You CF, Chu CH (1997) Tool-path verification in five-axis machining of sculptured surfaces. Int J Adv Manuf Technol 13(4):248–255

Study of Selective Laser Melting Process Parameters to Improve the Obtainable Roughness of AlSi10Mg Parts

Luana Bottini

Abstract Selective Laser Melting of AlSi10Mg parts has a lot of applications in different fields such as aerospace and automotive for its abilities to fabricate components characterized by complex shapes, good mechanical properties and low porosity. One of the main drawbacks for its application is the obtainable surface roughness widely not suitable for functional requirements. Typically, the improvement is handled by secondary operations thus markedly increasing the production time and costs. In this paper the possibility to improve the surface roughness by tuning Selective Laser Melting process parameters is investigated. A design of experiments is carried out considering not only the common laser process parameters but also the changing of the contour, upskin and downskin strategy definitions. This way the attained results show a marked decreasing of the roughness for vertical and horizontal surfaces.

Keywords Selective laser melting · AlSi10Mg · Roughness

Nomenclature

SLM	Selective Laser Melting
AM	Additive Manufacturing
b_o	Beam offset
h_o	Contour offset
s_o	Overlapping between lines
h_d	Hatch distance
P	Laser power

L. Bottini (✉)
Departement of Mechanical and Aerospace Engineering, Sapienza University of Rome, via Eudossiana 18, 00184 Rome, Italy
e-mail: luana.bottini@uniroma1.it

E. Ceretti and T. Tolio (eds.), *Selected Topics in Manufacturing*,
Lecture Notes in Mechanical Engineering,
https://doi.org/10.1007/978-3-030-57729-2_9

L	Layer thickness
v	Scan speed
VED	Volumetric Energy Density
LED	Linear Energy Density
R_a	Average roughness
R_t	Peak to valley roughness
R_{sk}	Skewness of profile height
R_{ku}	Kurtosis of profile height
Δ_q	Root mean square of the profile slopes
λ_c	Cut-off
λ_s	Wavelength cut-off

1 Introduction

Selective Laser Melting (SLM) is an Additive Manufacturing (AM) technology that belongs to the category of the power bed fusion [1]: it uses thermal energy provided by a laser source to selectively melts regions of a powder bed. This way it is possible to fabricate layer by layer full dense metallic components characterized by very complex geometries and mechanical properties comparable to those of bulk materials produced by traditional technologies [2]. One of the most used material is the aluminum that has the potential for applications and developments in different fields such as aviation, aerospace, automotive, naval and power electronics due to its low density, high specific strength and good corrosion resistance as well as its excellent electric and thermal conductivity [3–5].

Notwithstanding its industrial diffusion, many issues exist in the SLM processing of aluminum alloys due to their physical properties. The spreading of the powders in thin layers is difficult due to their low density and poor flowability leading to the formation of holes in the powder bed especially in presence of humidity [6]. The low absorptivity at laser wavelength and the high reflectivity require high specific energy [7]. The high thermal conductivity causes a fast cooling and a heat dissipation away from the melt pool [8]. The stirring and the incorporation of the powder oxide film into the melt pool can cause defects inside the part. The oxidized surfaces reduce the wettability of the built part and a high power is necessary to disrupt this oxide allowing the bind of the next layers and the realization of dense components [6].

The quality of SLMed parts depends upon the selection of the process parameters such as the laser power, the scanning speed, the hatch spacing, the layer thickness and the built orientation. The use of improper process parameters can cause defects such as cracks, low density, balling, satellite and dropping [9]. These surface defects deeply affect the roughness that typically is too high to meet the demand of the industrial production. In [10] the authors studied the influence on the surface roughness of the building orientation by the development of a roughness prediction model as a function of the local stratification angle. It highlighted the deep difference of morphology for surfaces characterized by different building orientations. In [11] the effects of the surface slope with different angles were studied by simulation and experimentation: as the slope increased, the part quality improved but the dimensional deviation worsened.

Calignano et al. [12] used a Taguchi method to study the effect of the laser power, the scan speed and the hatching spacing on surface roughness of horizontal surfaces of AlSi10Mg SLMed parts. They found that the scanning speed has the greatest influence on the surface quality. In [13] the influence of the same process parameters was altered such that better surface roughness for horizontal surfaces could be achieved for AlSi10Mg parts. The results showed that the best outcome was obtained for high specific energy at the lowest experimented beam offset. With this process parameters set also the minimum porosity inside the specimen was reached. In [14] only vertical surfaces were considered: the effect of the laser power, the scanning speed and the linear energy density on the morphologies of single tracks and on the roughness of vertical surfaces of cubic specimens were studied. They found a reduction more than the 70% of the average roughness for energy density in the range of 4.5–7.4 J/cm.

Another chance to reduce the roughness and satisfy functional requirements is the post processing such as surface treatments and machining. Common operations are sand blasting, machining, chemical etching, and plasma spraying but they are skill operator-dependent, labor-intensive, and difficult to apply to complex shape parts [15]. Although the post processing operations improve the surface quality, an increasing of production time and costs is required thus weakening most of the advantages of the SLM process in term of flexibility, efficiency and direct fabrication [10].

A compromise is the use of laser remelting: multiple scanning strategies can be performed during the SLM process on the external surfaces or on the entire layer. In [16] the influence of laser remelting on density, surface quality and staircase effect of AISI 316L SLMed parts was studied varying the operating parameters such as scan speed, laser power and hatch spacing. Moreover, a comparison between the remelting during and after the SLM process for inclined surfaces was done using the same machine. The results showed that when the laser remelting was applied only to the contour during the SLM process the staircase effect was reduced by 10–15%; conversely the remelting after the SLM fabrication provided a reduction of 70% in the average roughness. Notwithstanding the better result of the second strategy, it requires longer time to plan the remelting operation and it is limited to simple convex and accessible shapes. In [17] the remelting of the entire layers of AlSi10Mg part was performed in order to study the variation of the surface roughness of horizontal surfaces, the microstructure, the microhardness, the characteristic of the melt pool and the relative density. The results showed that the laser remelting enhanced the surface roughness and the relative density; it permitted to obtain a shallower melt pool, finer microstructure and higher hardness but it implied longer building time. In [18] a method for the correction of the porosity by remelting was studied: three different scan strategies were considered and experimented. Also, the surface roughness, the geometrical accuracy and material microhardness were evaluated. The results showed that the polishing strategy was the best solution to improve the porosity and to obtain smooth surfaces free of pits and protruded zones. Vaithilingam et al. [19] investigated the effect on surface chemistry of a double scanning only on the skin of Ti6Al4V components fabricated through SLM: they found that it altered the chemical composition leading to a significant reduction of the corrosion resistant and the biocompatibility but it permitted an improving of the surface quality. This method was effective on horizontal surfaces while inclined ones were not subjected to a significant improvement.

Aim of this work is to improve the obtainable surface roughness of AlSi10Mg parts characterized by surfaces with different inclinations performing a finishing operation directly inside the SLM process. For the purpose a remelting was performed for each layer contour after its fabrication with standard process parameters: the remelting strategy regarded only the external skin, this way the building time is slightly affected by the remelting. Vertical, horizontal and inclined surfaces required different scan strategies thus different remelting approaches were designed. An experimental plane was performed in order to find suitable process parameters for the different scan strategies.

2 Material and Methods

2.1 Machine and Material

The employed machine is an EOSINT®M290 located in the AM Lab of Sapienza University of Rome. It is equipped with a building platform of 250 × 250 × 325 mm^3 and a 400 W Ytterbium fiber continuum laser characterized by a beam spot size of 100 μm. Samples were fabricated using AlSi10Mg gas atomized powder supplied by EOS. It presents the nominal chemical composition reported in Table 1.

2.2 Scan Strategy and Process Parameters Definitions

During the SLM process, the laser beam moves over the surface of the powder bed in order to consolidate a layer. The scanning strategy is typically divided into two types: the part hatching and the part contour. The former is the processing of the internal area of the layer providing part mechanical resistance. The latter is developed according to the layer boundary and allows for better surface quality. A specific laser path must be generated considering geometrical elements and offsets. In Fig. 1 a schematization of the laser path definition is reported.

During the laser scan, namely the exposure, a consolidation zone of the solidified metal forms around the laser beam. The size of this zone depends upon the material used and the exposure parameters set. A beam offset b_o is introduced to compensate for the consolidation zone. The contour part is moved by this value with respect to

Table 1 Chemical composition of the AlSi10Mg powders

Element	Si	Fe	Cu	Mn	Mg	Ni	Zn	Pb	Sn	Ti	Al
Weight (%)	9.0 ÷ 11	≤0.55	≤0.05	≤0.45	0.2 ÷ 0.45	≤0.05	≤0.10	≤0.05	≤0.05	≤0.15	Balance

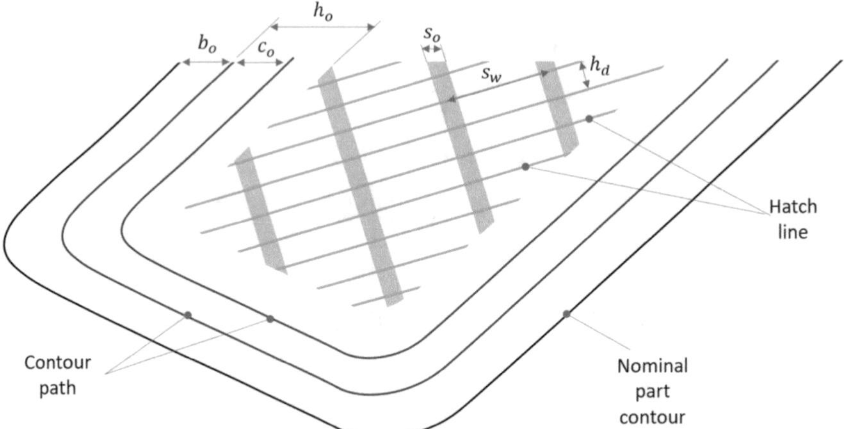

Fig. 1 Part layer, contour path, hatch lines schematization

the nominal part contour. Generally, more than one contours are generated moving furtherly the path. This parameter is called contour offset h_o. The filling of the part is provided by the hatching which moves the laser beam along parallel paths, namely the hatch lines, with properly defined energy. The interface between contours and hatching area is tailored by setting the hatch offset which defines the limit of the enclosed area. The hatch lines are organized in an exposure pattern. In the case of AlSi10Mg the hatching is exposed in stripes. In order to reduce the in-layer residual stress, a maximum hatch line length, namely the stripe width s_w, is set. The obtained discontinuity is reduced by an overlapping between lines called s_o. The distance between the lines is called hatch distance h_d.

This parameter, together the laser power P and the layer thickness L, is involved in the well-known formula of Volumetric Energy Density (VED):

$$VED = \frac{P}{Lvh_d} \tag{1}$$

where v is the scan speed. For the contour Eq. (1) cannot be applied thus the Linear Energy Density (LED) is used:

$$LED = \frac{P}{v} \tag{2}$$

The hatching of the top and the bottom of the part is no longer inside the part itself thus requiring different exposure settings. For the purpose the infill is divided into three categories: the upskin (zone over which no area to be exposed is present), the infill (interior area) and the downskin (zone under which no area to be exposed is present). The upskin and downskin thickness is defined by the number of layers assigned to these categories. An overlap between infill and upskin/downskin is assigned to cover the discontinuity. In Fig. 2 a representation of these zones is shown.

In this work a laser remelting of the part skin was performed. The consolidation of the specimen was obtained using EOS standard process parameters for AlSi10Mg. The used layer thickness was 30 μm and hatching and contour had the processing parameters set reported in Table 2. As shown the infill, the upskin and the downskin are processed at about the same VED, i.e. 50, 57 and 47 J/mm^3 respectively pair to a LED of about 0.3 J/mm. Conversely the contour, which is composed by two similar scans, is characterized by a very small energy: the LED is reduced to 0.09 J/mm.

Aim of this work is to investigate how to improve the surface roughness of the part by providing a skin laser remelting. A typical approach to consider factors' level is the VED and LED for upskin and contour respectively. In [12] a fractional experimental plan was done varying scan speed between 800 and 1250 mm/s, setting the hatch distance at 0.15 and 0.20 mm and the power in the range 120–190 W (limited by the maximum system value); the resulting VED was in the range 24–79 J/mm^3. In [13] authors employed a greater laser power and investigated process outcome accordingly to manufacturer's standard parameters: the correspondent VEDs were: 50, 57, 133 J/mm^3. In the present work a wide range of VED for upskin is investigated: varying processing parameters at the three levels reported in Table 3, the VED ranges between 28 and 167 J/mm^3. Figure 3a shows the slice contour plot of VED: it is well evident that the considered factors levels allow having distinct energy

Fig. 2 Part vertical section: contour (C), upskin (U), overlap (O), downskin (D) and infill (IN) zones

Table 2 Process parameters for the consolidation of the part

Hatch				Contour	
Infill		Upskin		1	
P (W)	370	P (W)	360	P (W)	80
V (mm/s)	1300	V (mm/s)	1000	V (mm/s)	900
Hd (mm)	0.19	Hd (mm)	0.21	Co (mm)	0.02
Ho (mm)	0.02	Downskin		2	
Sv (mm)	0.02	P (W)	340	P (W)	85
Sw (mm)	7	V (mm/s)	1150	V (mm/s)	900
So (mm)	0.02	Hd (mm)	0.21	Co (mm)	0

Table 3 Experimented process parameters for upskin and contour strategy

Upskin				Contour			
Factors	Levels			Levels			
P (W)	240	300	360	P (W)	150	250	350
V (mm/s)	400	800	1200	V (mm/s)	400	800	1200
Hd (mm)	0.18	0.21	0.24	Co (mm)	0	0.02	0.04

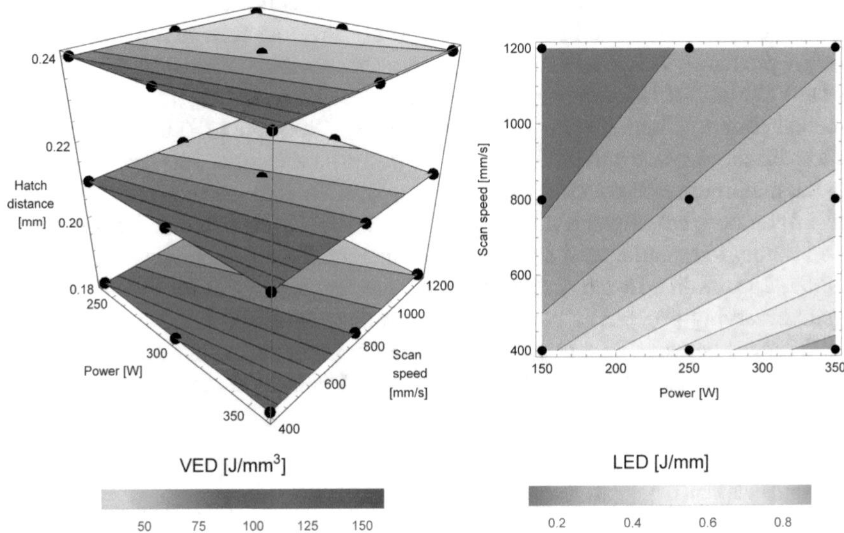

Fig. 3 VED (**a**) and LED (**b**) as a function of process parameters

values. As regards the contour, the work [14] reports several experimentations on vertical surfaces suggesting a LED in the range 0.46–0.74 J/mm: at higher energy, i.e. 0.82 J/mm, the Marangoni convection is enough strong to disturb the molten pool and fluctuations appear. Thus, in the present experimentation the range 0.125–0.875 J/mm was chosen. The relative process parameters are reported in Table 3 and represented in Fig. 3b.

The designed specimen and the fabricated ones are reported in Fig. 4. The geometry is characterized by horizontal, vertical and two inclined surfaces, 45 and 135°. The support structures were avoided on 135° overhanging surface and solid strategy was applied to the underside of the specimen.

The building platform was kept at 160 °C during the fabrication process in order to reduce residual stresses and no laser conditioning was applied to infill areas thus maintaining bulk properties of the fabricated specimen. For the purpose also the between-layer hatch strategy, such as the hatch rotation of 67°, was left unchanged. The specimens were oriented by 5° and fuzzy placed onto the platform so that the recoater concurrent hits were reduced; the exposure order was set against the argon

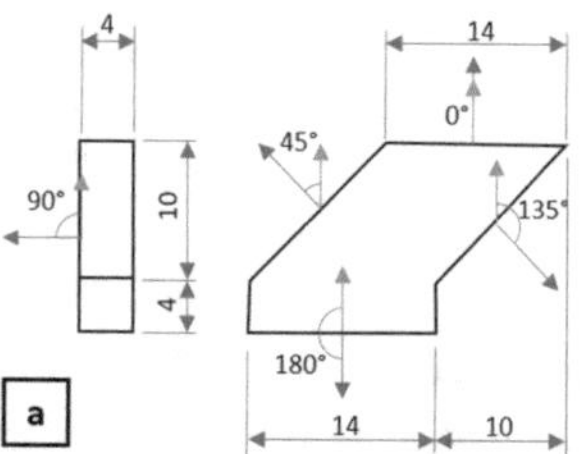

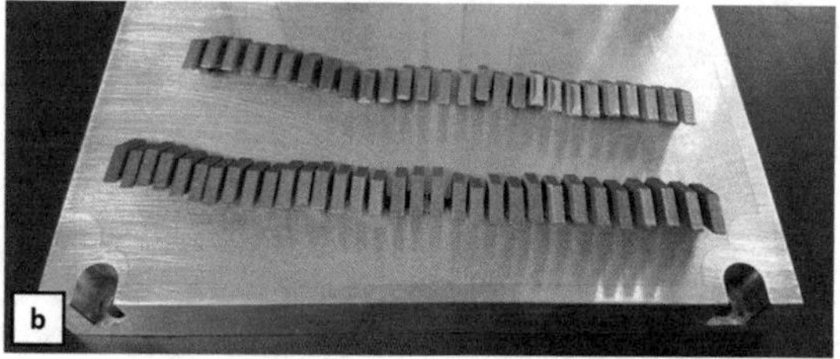

Fig. 4 Designed specimen (**a**) and fabricated specimens (**b**)

flow to avoid processing smoke onto powder surfaces going to be scanned. A replication of each parameter set combination was provided so that a total of 54 specimens were built. In addition, some specimens were fabricated without remelting for comparison. After the fabrication a thermal treatment for 2 h at 300 °C was performed in order to reduce deformation during the detachment from the building platform. This cutting operation was provided by an abrasive metallographic cutting machine: a SiC abrasive wheel 400 mm in diameter was used at 40 m/s cutting speed and 20 mm/min feed speed. No finishing operation, such as shot peening, was applied in order to leave the original SLMed surfaces.

2.3 *Surface Measurements and Morphological Analysis*

The roughness measurements were performed by a Mitutoyo Surftest SJ-412: the sampling length and the evaluation length were set at 2.5 and 12.5 mm, respectively [20]. A 2-μm stylus, sampling every 1.5 μm, was employed. The data were processed by a spline profile filter [21] with a cut-off λ_c and a short wavelength cut-off λ_s equal to 2.5 mm and 8 μm, respectively. The roughness profile parameters were calculated according to [22].

The morphological analysis was provided through an optical microscope. Each specimen was sectioned by a Struers Labotom-5 abrasive cutting equipment and polished by a Struers Labopol-2. The sections and the surfaces were captured by a Dino-Lite digital microscope with a polarized filter to remove the unwanted reflection or glare from the shiny object surface. An extension of the depth of field was obtained by moving the object along the optical axis and restored by wavelet-based image fusion technique [23].

3 Results

3.1 Upskin

The upskin strategy affects the surface quality of horizontal surfaces. According to the prediction model reported in [10], the attainable average roughness is about 7 μm. The specimen without remelting, i.e. a specimen fabricated with the standard processing parameters, was analyzed by means of roughness measurements (Fig. 5a). The average roughness is 7.57 μm very close to the expected value, and the total roughness R_t is more than 8 times this value highlighting the presence of local high peaks. In fact, the amplitude density function of the profile heights is characterized by pronounced tails; moreover, around the mean line a large spread is observed. This distribution is symmetrical (R_{sk} about zero) and the peakedness is leptokurtic (R_{ku} equal to 3.43). The mean spacing is 0.26 mm. This profile is characterized by a low reflectiveness as quantified by the Δ_q pair to 0.29, i.e. about 16°.

As a remelting is applied with a high VED (Fig. 5b) the roughness is improved: the defects are reduced as confirmed by an R_a pair to 2.8 μm and a R_t less than 6 times greater; the reflectiveness is improved to a Δ_q pair to 0.05 that is about 3°. The height distribution is now slightly platykurtic, the mean spacing is 0.57 mm. At 1200 mm/s scan speed (Fig. 5c) some local peaks are observed leading to an intermediate result for the average roughness and the reflectiveness. By increasing the VED at 139 J/mm^3 (with the process parameters set reported in Fig. 5d) a 76% decreasing in R_a with respect to the as is specimen is obtained. As expected, if the

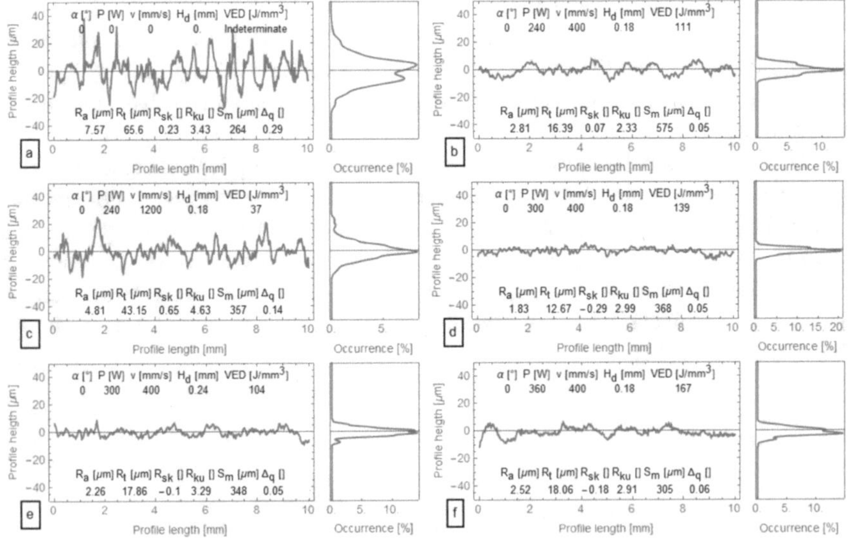

Fig. 5 Roughness profiles and amplitude density functions for different upskin remelting conditions

hatch distance is increased (VED pair to 104 J/mm^3), a reduced improvement is provided (Fig. 5e). As the VED is increased by a power of 360 W (Fig. 5f) the R_a is 2.52 μm suggesting that the decreasing in R_a is not linearly dependent upon the VED. The main effect plot can help in this investigation (Fig. 6a). As expected, the increasing of the power can lead to a decreasing of the R_a even if the effects at 300 and 360 W are the same; a marked increase is obtained by increasing the speed; conversely the hatch distance is slightly affecting the R_a. Both the VED and LED show a relationship with the R_a. In particular a local minimum is obtained at about 140 J/mm^3. In Fig. 6b, the box plot for different processing parameters values is displayed. It can be assessed that the best condition is found at 300 W power, 400 mm/s scan speed and 0.18 mm hatch distance but also applying 360 W at 400 and 800 mm/s a relevant improving is achieved.

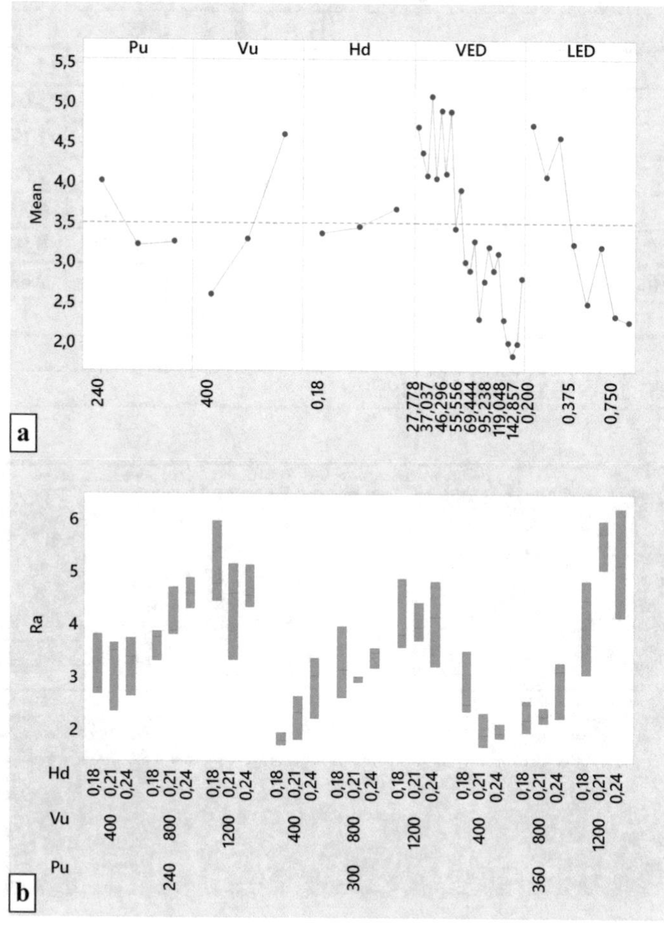

Fig. 6 Main effect plot (**a**) and box plot (**b**) for the experimented upskin process parameters

In the metallographic analysis the presence of defects will address the choice far from the highest VED levels. The analysis of variance (ANOVA) reported in Table 4 highlights that the hatch distance is not significant while the power, the scan speed and their interaction significantly affect the R_a (p-value < 0.01).

Fabricated surfaces were analyzed by means of morphological behavior. In Fig. 7a the specimen not subjected to remelting is shown. As expected, the presence of balling and satellite is observed. As a little specific energy is applied these defects decrease in dimension and occurrence (Fig. 7b). The laser tracks are herewith well evident but instable causing big pores: this is probably due to the strong Marangoni convection. At about 60 J/mm^3 the roughness improvement is accompanied by sparse balling (Fig. 7b) and unexpected thermal cracks at large hatch spacing (Fig. 7c). The number

Table 4 ANOVA table for the upskin data

Source	DF	SS	MS	F	P
Pu	2	11.079	5.5395	17.08	0.000
Vu	2	55.760	27.8801	85.94	0.000
Hd	2	1.407	0.7037	2.17	0.124
Pu * Vu	4	9.934	2.4836	7.66	0.000
Pu * Hd	4	0.305	0.0762	0.23	0.918
Vu * Hd	4	0.896	0.2239	0.69	0.602
Pu * Vu * Hd	8	6.955	0.8694	2.68	0.015
Error	54	17.517	0.3244		
Total	80	103.854			
S = 0.569557, R^2 = 83.13%, adj R^2 = 75.01%					

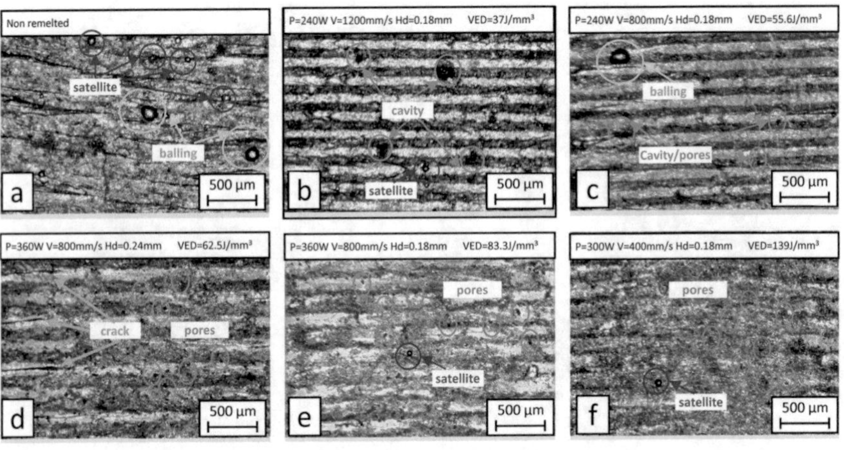

Fig. 7 Morphologies of horizontal surfaces without remelting (**a**) and at different remelting conditions (**b–f**)

of pores, probably due to shrinkage, increases. The adhering particles are reduced and laser tracks are smooth. Thermal cracks and almost all the defects disappear as the VED increases to 83 J/mm^3 (Fig. 7d) and 139 J/mm^3.

3.2 Contours

The vertical and the inclined surfaces are typically characterized by a very rough surface. According to the model [10], the vertical one is characterized by 22 μm R_a: this is confirmed by the profile reported in Fig. 8a: the distribution is symmetrical and Gaussian (R_{sk} about 0 and R_{ku} about 3) indicating a chaotic behavior of the profile heights. If a small LED is applied (Fig. 8b) the surface is slightly affected by the conditioning: the R_a is about unchanged, the distribution is leptokurtic and asymmetric since it moves downwards indicating an emptied profile. Negligible changing is observed if contour offset is modified. If the LED is increased by increasing power, the profile slightly worsens as reported in Fig. 8c. At this particular linear energy, if the contour offset is set to zero, an impressive changing is observed: the R_a is 4.2 μm and the peak to valley height is more than one quarter of millimeter (Fig. 8d). When the LED is more than 0.4 J/mm a marked improving in surface roughness is obtained. The process conditions reported in Fig. 8e provide an R_a under 5 μm and no variation is observed at different contour offset. At higher LED levels negligible differences are experienced as shown in Fig. 8f. The ANOVA analysis (Table 5) confirms this

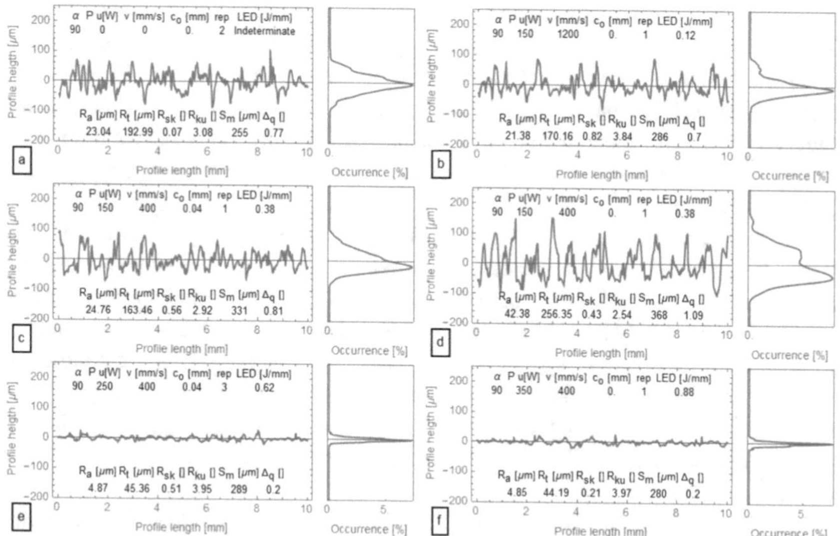

Fig. 8 Roughness profiles and amplitude density functions for different contour remelting conditions

Table 5 ANOVA table for the contour data

Source	DF	SS	MS	F	P
Pc	2	2616.47	1308.24	168.95	0.000
Vc	2	698.70	349.35	45.12	0.000
Co	2	70.06	35.03	4.52	0.015
Pc * Vc	4	2540.54	635.14	82.03	0.000
Pc * Co	4	364.95	91.24	11.78	0.000
Vc * Co	4	106.61	26.65	3.44	0.014
Pc * Vc * Co	8	70.92	8.86	1.14	0.349
Error	54	418.13	7.74		
Total	80	6886.39			
S = 2.7826, $R^2 = 93.93\%$, adj $R^2 = 91.00\%$					

intricate behavior assessing that the R_a is significantly affected by the factors and most of their interactions.

From the main effect plots (Fig. 9a) the expected trends for processing parameters are observed. The LED confirms a wave trends with a marked decreasing at high values. The box plots (Fig. 9b) suggest three scenarios corresponding to more than 70% roughness reduction: at {350 W, 400 mm/s, 0.04 mm}, at {250 W, 800 mm/s, 0.04 mm}, at {350 W, 800 mm/s, 0 mm}.

The untreated vertical surfaces are characterized by a generalized presence of balling as evidenced in Fig. 10a. The Marangoni convection here is so strong that the application of a mild linear energy, i.e. 0.125 J/mm, does not take effect (Fig. 10b). If a LED of 0.375 J/mm is applied the energy causes instabilities and generates local structures of balling coming from different layers (Fig. 10c). When the linear energy exceeds a threshold, the stability is no longer influenced by the LED and surface quality is markedly increased. Figure 10d shows the outcome at maximum power, intermediate scanning speed and contour offset pair to zero, which corresponds to a LED of 0.625 J/mm.

The same analysis can be carried out for surfaces inclined by 45 and 135°. The former is characterized by a 23 μm R_a in the non-treated case. This slope is markedly affected by the upper side of the contour and in most of the considered parameters set the remelting is not effective like in the vertical and horizontal surfaces. As evidenced in the main effect plot of Fig. 11a, the behavior with main processing parameters is not expected and ANOVA assesses the power and the speed are not significantly affecting the outcome. As regards the LED, two scenarios can be suggested: {250 W, 400 mm/s, 0 mm} and {350 W, 800 mm/s, 0 mm}. In these conditions the R_a is reduced to 16.5 μm.

In the case of 135° the starting R_a is 47 μm. Also in this case the ANOVA highlights an heterogeneous behaviour (Fig. 11b) and the LED can be used to find a minimum: the suggested scenario is: {250 W, 800 mm/s, 0 mm} with a reduction of 24%.

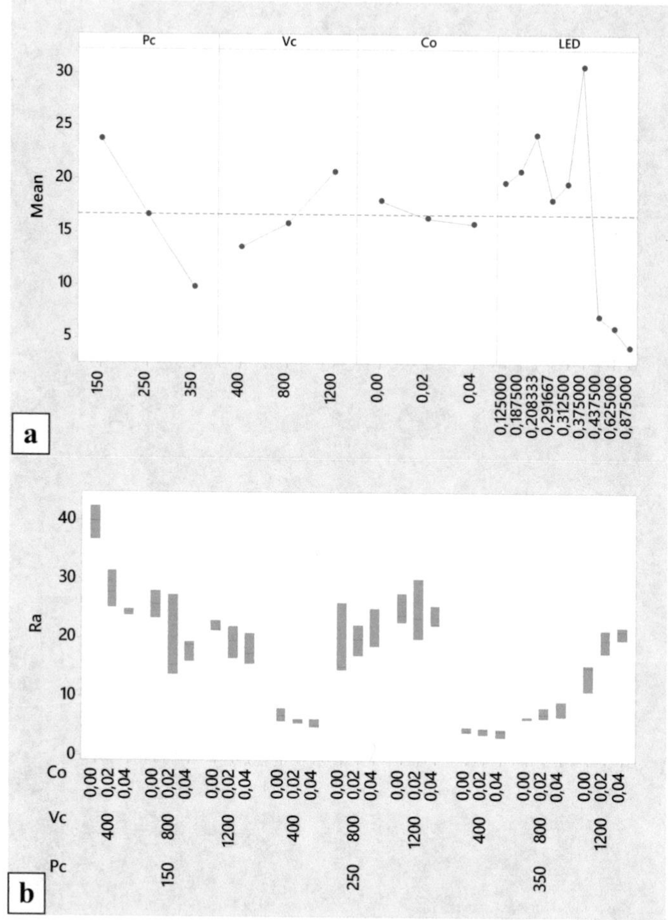

Fig. 9 Main effect plot for vertical surfaces (**a**) and box plot (**b**) for the experimented contour process parameters

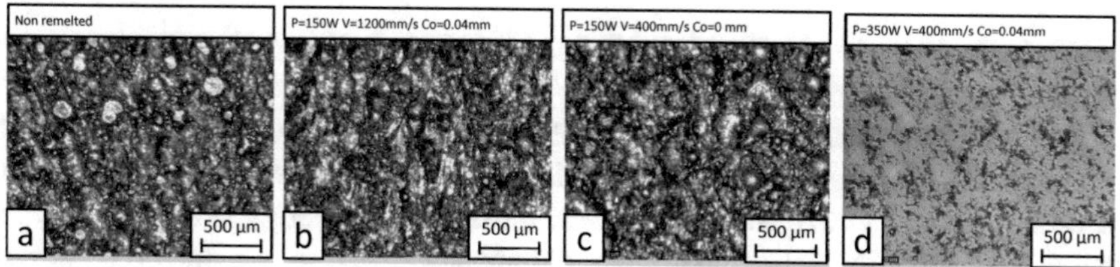

Fig. 10 Morphologies of vertical surfaces without remelting (**a**) and at different remelting conditions (**b–d**)

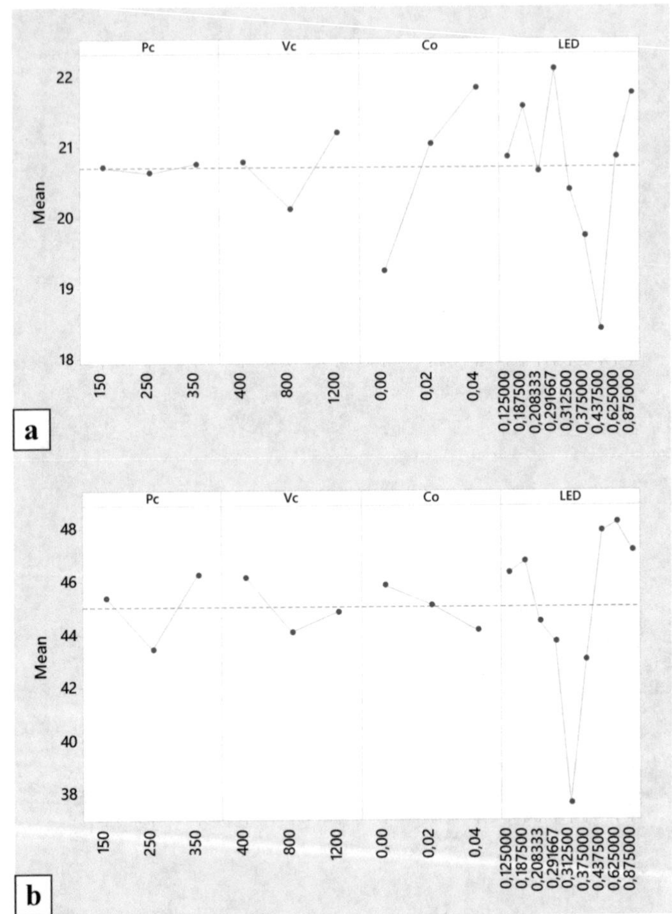

Fig. 11 Main effect plots of the experimented contour process parameters for surfaces characterized by 45° (**a**) and 135° (**b**)

The morphological analysis confirms the marginal improvement of selected processing parameters. In Fig. 12a the not remelted surface is fully characterized by balling phenomena. At a LED pair to 0.62 J/mm a scattering of defects is observed (Fig. 12b). For overnaging surfaces the peaks are twice the previous case and cannot be detected on Fig. 12c. This surface needs less specific energy, i.e. 0.31 J/mm, to find a good combination of processing parameters which may reduce the balling phenomena.

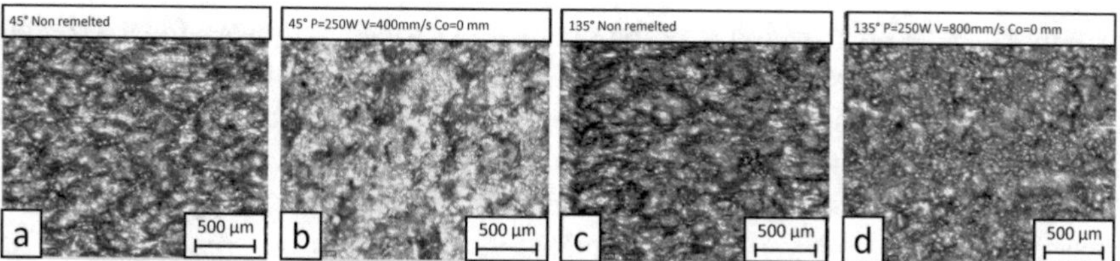

Fig. 12 Morphologies of inclined surfaces without remelting (**a**, **c**) and with remelting (**b**, **d**)

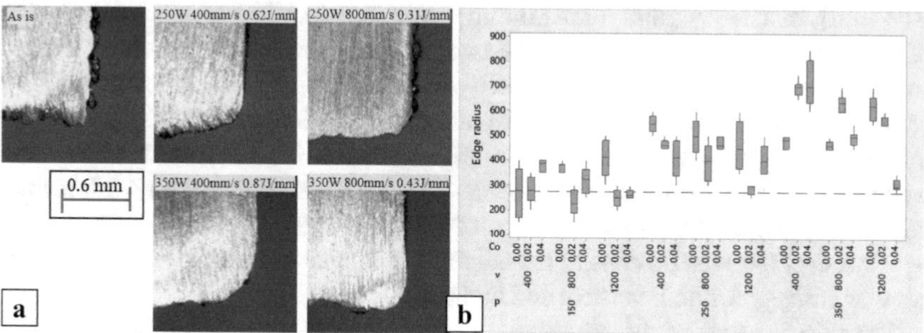

Fig. 13 Fabricated edges for different experimented conditions (**a**); box plot of edge radii (μm) (**b**)

3.3 Edge Deviation

The remelting leads to a modification of micro-geometry which may affect also the shape of the specimen. This is well evident if the edges are considered. A not treated specimen shows a mean edge radius of 0.275 mm (Fig. 13a). This deviation is typical of aluminum alloys which present the issues claimed in the introduction. When the LED is low this deviation is almost maintained (less than 0.5 J/mm). If the remelting is carried at 0.87 J/mm (350 W and 400 mm/s) a big radiusing occurs: in this condition 0.7–0.8 mm is measured and it must be checked if it is tolerated by the part design. The LED is not a reliable way to understand this behavior: an exception occurs at middle value (350 W, 1200 mm/s, 0.04 mm) where the measured radius ranges between 0.6 mm and 0.7 mm. The undertaken measures are reported in the box plot of Fig. 13b. It is well evident that the power must be kept at 250 W and some of the previous scenarios are preferred.

4 Conclusions

In this paper the effects of a skin laser remelting of AlSi10Mg parts fabricated by SLM was studied. Different process parameters for different scan strategies were experimented in order to improve the obtainable surface roughness for horizontal, vertical and inclined surfaces. The upskin strategy influenced the quality of the horizontal ones. The best result for the roughness was obtained for the parameters set 300 W power, 400 mm/s scan speed and 0.18 mm hatch spacing with a reduction more than 75% with respect to the as is specimen. The contour strategy influenced the surface quality of both vertical and inclined surfaces. For these surfaces, different behaviors at different experimented process parameters were observed. The best surface quality for the vertical surfaces were attained at higher LED values. The suggested parameters set were {350 W, 400 mm/s, 0.04 mm}, {250 W, 800 mm/s, 0.04 mm}, at {350 W, 800 mm/s, 0 mm} corresponding to a roughness reduction of more than 70%. Conversely for the two inclined surfaces the remelting was not so effective. For 45°, two scenarios were suggested {250 W, 400 mm/s, 0 mm} and {350 W, 800 mm/s, 0 mm} that correspond to a reduction about 28%; for 135° the minimum roughness was found for the parameters set {250 W, 800 mm/s, 0 mm} with a reduction of 24% than the as is case. The shape was analyzed highlighting that high energy values determine defects on the edges shape. To reduce these effects a power reduction must be considered helping the choice of suitable scenarios of the contour strategy for both vertical and inclined surfaces.

The effectiveness of the proposed remelting methodology is accompanied by a small increase of building time: the simulation over the well-known NIST benchmark artefact [24] required only 2% additional building time with respect to the fabrication without remelting.

Further investigations will regard the application of different process parameters set to all the possible inclinations.

References

1. ISO/ASTM52900-15 (2015) Standard terminology for additive manufacturing—general principles—terminology. ASTM International, West Conshohocken
2. Gu D (2015) Laser additive manufacturing of high-performance materials. Springer-Verlag, Berlin Heidelberg
3. Martin JH, Yahata BD, Hundley JM, Mayer JA, Schaedler TA, Pollock TM (2017) 3D printing of high-strength aluminium alloys. Nature 549:365–369
4. Jung JG, Lee SH, Cho YH, Yoon WH, Ahn TY, Ahn YS, Lee JM (2017) Effect of transition elements on the microstructure and tensile properties of Al–12Si alloy cast under ultrasonic melt treatment. J Alloys Compd 712:277–287
5. Jung JG, Ahn TY, Cho YH, Kim SH, Lee JM (2018) Synergistic effect of ultrasonic melt treatment and fast cooling on the refinement of primary Si in a hypereutectic Al–Si alloy. Acta Mater 144:31–40
6. Louvis E, Fox P, Sutcliffe CJ (2011) Selective laser melting of aluminium components. J Mater Process Technol 211:275–284
7. Gusarov AV, Kruth JP (2005) Modelling of radiation transfer in metallic powders at laser treatment. Int J Heat Mass Transfer 48(16):3423–3434
8. Fischer P, Karapatis N, Romano V, Glardon R, Weber HP (2002) A model for the interaction of near-infrared laser pulses with metal powders in selective laser sintering. Appl Phys A Mater 74(4):467–474

9. Zhang J, Song B, Wei Q, Bourell D, Shi Y (2019) A review of selective laser melting of aluminum alloys: processing, microstructure, property and developing trends. J Mater Sci Technol 35(2):270–284
10. Boschetto A, Bottini L, Veniali F (2017) Roughness modeling of AlSi10Mg parts fabricated by selective laser melting. J Mater Process Technol 241:154–163
11. Xiang Z, Wang L, Yang C, Yin M, Yin G (2019) Analysis of the quality of slope surface in selective laser melting process by simulation and experiments. Optik 176:68–77
12. Calignano F, Manfredi D, Ambrosio EP, Iuliano L, Fino P (2013) Influence of process parameters on surface roughness of aluminum parts produced by DMLS. Int J Adv Manuf Technol 67:2743–2751
13. Mohammadi M, Asgari H (2018) Achieving low surface roughness AlSi10Mg_200C parts using direct metal laser sintering. Addit Manuf 20:23–32
14. Yang T, Liu T, Liao W, MacDonald E, Wei H, Chen X, Jiang L (2019) The influence of process parameters on vertical surface roughness of the AlSi10Mg parts fabricated by selective laser melting. J Mater Process Technol 266:26–36
15. Boschetto A, Bottini L, Veniali F (2018) Surface roughness and radiusing of Ti6Al4V selective laser melting-manufactured parts conditioned by barrel finishing. Int J Adv Manuf Technol 94:2773
16. Yasa E, Kruth J (2011) Application of laser re-melting on selective laser melting parts. Adv Prod Eng Manag 6:259–270
17. Liu B, Li BQ, Li Z (2019) Selective laser remelting of an additive layer manufacturing process on AlSi10Mg. Results Phys 12:982–988
18. Demir AG, Previtali B (2017) Investigation of remelting and preheating in SLM of 18Ni300 maraging steel as corrective and preventive measures for porosity reduction. Int J Adv Manuf Technol 93:2697–2709
19. Vaithilingam J, Goodridge RD, Hague RJM, Christie SDR, Edmondson S (2016) The effect of laser remelting on the surface chemistry of Ti6Al4V components fabricated by selective laser melting. J Mater Process Technol 232:1–8
20. ISO 4288-08 (E) (1996) Geometrical product specifications (GPS)—surface texture: profile method—rules and procedures for the assessment of surface texture. International Organization for Standardization (ISO), Geneva
21. ISO 16610-22 (2015) Geometrical product specifications (GPS)—filtration—part 22: linear profile filters: spline filters. International Organization for Standardization (ISO), Geneva
22. ISO 4287 (1997) Geometrical product specification (GPS)—surface texture: profile method—terms, definition and surface texture parameters. International Organization for Standardization (ISO), Geneva
23. Forster B, Van De Ville D, Berent J, Sage D, Unser M (2004) Complex wavelets for extended depth-of-field: a new method for the fusion of multichannel microscopy images. Microsc Res Technol 65:33–42
24. Moylan, SP, Slotwinski JA, Cooke AL, Jurrens KK, Donmez MA (2012) Proposal for a standardized test artifact for additive manufacturing machines and processes. In: Proceedings of the 23rd international solid free form symposium—an additive manufacturing conference, Austin, Aug 2012, pp 902–920

Surface Modifications Induced by Roller Burnishing of *Ti6Al4V* Under Different Cooling/Lubrication Conditions

Giovanna Rotella and Luigino Filice

Abstract The paper presents a deep analysis of surface modifications induced by roller burnishing process of Ti6Al4V titanium alloy. The extensive experimental campaign has been performed based on a Design of Experiments at varying lubrication/cooling strategies (dry, cryogenic and MQL), roller radius, burnishing speed and burnishing depth. The resulting surface integrity has been analyzed in terms of surface roughness, micro hard-ness, microstructural changes and tribological performance. In particular, the wear rate of the burnished sample has been evaluated as a quality indicator of the process. The overall results show the influence of burnishing process parameters on surface quality and wear resistance of Ti6Al4V highlighting the capability of the process to significantly improve the above performance especially when cryogenic cooling is applied. Finally, the extensive experimental activity allowed to find a combination of processing parameters and lubrication conditions able to significantly improve the surface quality of the final component.

Keywords Burnishing · Surface integrity · Severe plastic deformation · Wear resistance

1 Introduction

Surface modification processes are often used to finish many components, especially those requiring high quality and service performance such as wear and corrosion

G. Rotella (✉)
Department of Informatics, Modeling, Electronics and Systems Engineering (DIMES), University of Calabria, 87036 Rende, CS, Italy
e-mail: giovanna.rotella@unical.it

L. Filice
Department of Mechanical, Energy and Management Engineering (DIMEG), University of Calabria, 87036 Rende, CS, Italy
e-mail: luigino.filice@unical.it

E. Ceretti and T. Tolio (eds.), *Selected Topics in Manufacturing*,
Lecture Notes in Mechanical Engineering,
https://doi.org/10.1007/978-3-030-57729-2_10

resistance, fatigue endurance etc. In fact, some manufacturing processes can alter the quality and geometric tolerances of the products requiring further post processing [1]. Thus, surface modification processes can be used to modify the properties of the original material limiting such changes on targeted areas of the components, typically the surface [2]. Therefore, different surface modifications can be performed according to the specific characteristics that each product needs to exhibit. Within these techniques, burnishing process is the one born to smooth the surface of machined components with interesting consequences like increase of fatigue strength and an overall enhanced surface integrity [3]. Burnishing is a cold working process in which the surface and subsurface layers' experience plastic deformation by a smooth rolling body pressed against the surface of the workpiece. The most used tools are spherical (ball burnishing) and cylindrical (roller burnishing). In the latter, the roller bodies need to be continuously in contact with the workpiece surface while rolling, thus, it is usually designed to be held and guided in a bearing system. The tool is then mounted on a milling machine or lathe and brought into contact with a machined surface with the aim to smooth the surface by applying a load greater than the yield strength of the component. Thus, the mirror like final surface and the overall superior surface integrity of the burnished product are achieved by varying a series of process parameters such as feed rate, burnishing depth, speed, roller geometry, number of passes, cooling/lubrication strategies, etc. [3–5]. Although studies have been conducted to understand the relationship within the burnishing parameters and the post processed surface integrity on many industrial materials [6–8], a clear knowledge when *Ti6Al4V* is burnished is still missing. Thus, it is crucial to perform deeper investigation aiming to assess the influence of process parameters on surface integrity of roller burnished *Ti6Al4V* and their effects on the overall wear resistance. In fact, the massive use of such alloy in the aerospace industry is mainly related to its combination of lightweight, strength, high temperature resistance and ductility. However, the poor tribological properties of titanium alloys represent still a limitation for their widespread use in many industrial fields. Thus, surface treatments of *Ti6Al4V* are commonly performed in order to improve its overall wear performance [9]. In this paper, the effects of burnishing depth, speed, cooling/lubrication techniques and tool radius on the surface integrity and wear resistance of the alloy are deeply studied by means of a full factorial Design of Experiments (DOE). The overall results offer a detailed operational framework of the burnishing processing influence on the final *Ti6Al4V* surface quality and product wear performance.

2 Materials and Methods

The material under investigation is the Grade 5 titanium alloy *Ti6Al4V*. The titanium bars (with a diameter of 30 mm) have been first finished by machining, subsequently, roller burnishing tests have been conducted using a high-speed CNC turning center equipped with a cooling line to supply different cooling and lubrication media. Burnishing tests have been performed using a custom-made roller-burnishing tool, made of a hardened steel fastened to a customized crotch. The burnishing tool was connected to a piezoelectric dynamometer for force detection. The bars were mounted on a lathe with a tailstock and their length and diameter were selected to be small enough to avoid deflection of the workpiece during burnishing. The temperatures reached during the tests have been monitored by an infrared thermo-camera and the experimental campaign has been carried out under dry, minimum quantity of lubrication (MQL) and cryogenic (cryo) lubrication/cooling conditions. The MQL tests have been performed applying vegetable oil through an external nozzle to the tool-workpiece contact zone with a flow rate of 60 ml/h while the cryogenic coolant (LN_2) has been delivered at the pressure of 6 bar through two nozzles (with an inner diameter of 2 mm) at a maximum distance to the contact region of 10 mm. It is worth pointing that the cryogenic liquid has been applied on the unprocessed surface during the process at a distance of 10 mm ahead the roller tool with a resulting pre freezing effect which avoids the problems encountered with the direct application of a conventional cryogenic process such as the risk to freeze the bearing systems holding the tool or reducing the flow capability of the material which usually lead to a poorer surface quality [4]. Figure 1 shows a schematic of the experimental equipment.

The full factorial experimental campaign has been carried out at varying burnishing speed (v), burnishing depth (a_p), roller radius (R) and lubricant conditions for a total of 81 different tests with three repetitions each. The parameters used during the experimental campaign are reported in Table 1. Furthermore, an explorative preliminary experimental campaign (27 tests) has been also performed in order to select the optimal feed rate and number of passes which have been fixed to 0.05 mm/rev and 2 respectively.

The range of tested number of passes and feed rate were 1, 2 and 3 passes and 0.05, 0.1 and 0.2 mm/rev respectively according to what stated in literature [3–5]. It is worth noting that the burnishing forces have been measured as an output of the

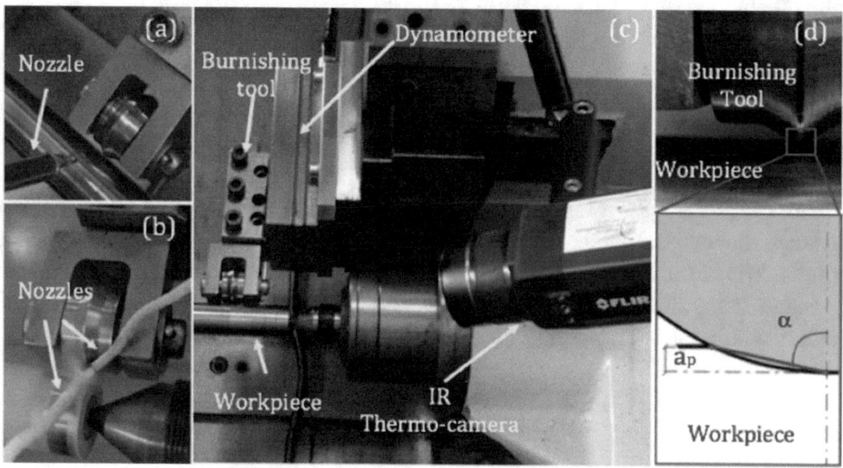

Fig. 1 Experimental setup for **a** MQL, **b** cryogenic and **c** dry burnishing; **d** burnishing tool details

Table 1 Design of experiments for the full factorial experimental campaign

Factor	Level 1	Level 2	Level 3
Lubricant/coolant	Dry	MQL	Cryogenic
Burnishing depth (mm)	0.1	0.3	0.5
Speed (m/min)	50	100	200
Radius (mm)	1	2.5	5

process instead of controlled input variables. This is due to the need to maintain the same burnishing depth and to guarantee the same dimensional tolerances in each of the investigated case. In fact, the proper setting of burnishing parameters is crucial to avoid exceeding the tolerance field. Moreover, it has been possible to customize the tool and the holder for the specific needs.

The cross section of burnished samples has been cut and mounted into a resin holder for further analysis. The specimens have been mechanically polished and etched using the Kroll's reagent (92 ml of distilled water, 6 ml of nitric acid, 2 ml of hydrochloric acid) and then analyzed using an optical microscope (1000×). Mean surface roughness *Ra* has also been measured by means of a non contact 3D confocal profilometer while the micro-hardness ($HV_{0.01}$) of the surface and subsurface layer has been probed by means of an instrumented micro-nano indenter.

The microhardness tests have been carried out through the depth of the burnished samples taking into account the edge effects. Thus, in order to avoid undesired results, the first micro indentation has been performed at a distance of about 20 μm from the burnished surface. The distance between consecutive indentations as well as that from the surface was always bigger than 3 times the longest diagonal of the impression. Another batch of burnished samples has been used to perform wear tests. In particular, a linearly reciprocating ball-on-flat sliding wear tests have been performed according to the ASTM G133 standard by means of a tribometer equipped with an alumina ball as a static partner. The tests were performed on the as received material, the burnished samples and the as machined surface.

The specific wear rate, was calculated using the Archard model reported in Eq. (1)

$$W_r = V/Fn * l \ \left(\mathrm{mm^3/N/m}\right) \tag{1}$$

where V is the sample volume loss, Fn is the average normal load and l is the sliding distance. A confocal white light 3D surface profilometer was used to measure the volume loss of the burnished specimen for the evaluation of wear rate.

Figure 2 reports a further detail of the roller geometry highlighting the volume contact within the tool and the roller at a burnishing depth of 0.3 mm.

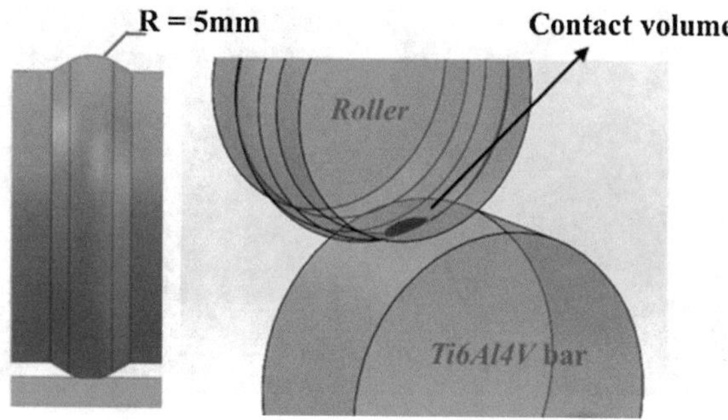

Fig. 2 Sketch of the roller and the contact with the workpiece

3 Results and Discussion

3.1 Burnishing Force

Figure 3 shows as the burnishing forces, defined as the force in the radial direction, are influenced by the investigated process parameters. The burnishing forces increase with increasing of burnishing depth since a larger localized cold worked zone is created leading to higher plastic deformation.

These results are in agreement with those observed by Luo et al. [10], where it was demonstrated that burnishing forces increase approximately exponentially with the increase of the burnishing depth. Also, larger tool radius generates increasing burnishing forces even if to a lower extent than those produced by varying the burnishing depth. Moreover, the forces increase under cryogenic burnishing due to the higher strain hardening capability of the investigated alloy [11, 12]. In contrast, lower burnishing forces for dry and MQL have been measured and both conditions

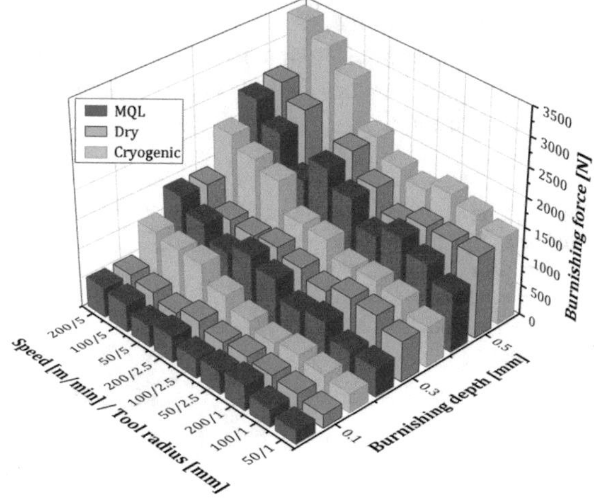

Fig. 3 Mean burnishing force measured at varying the process parameters, the tool geometry and the cooling/lubrication conditions

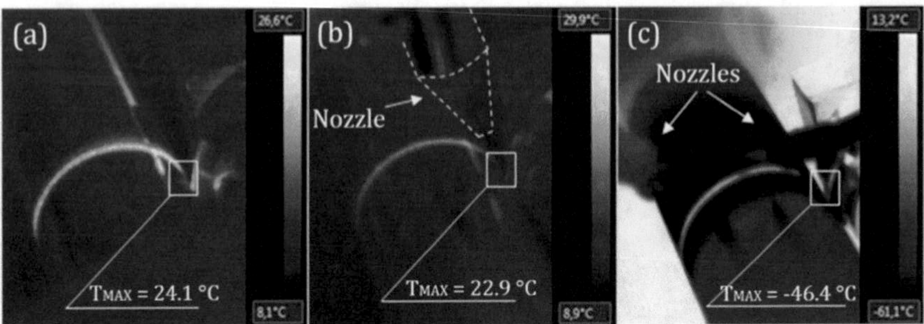

Fig. 4 Steady-state measured temperatures under **a** dry, **b** MQL and **c** cryogenic conditions for a speed of 200 m/min, tool radius of 5 mm and burnishing depth of 0.5 mm

display similar trends since they work in a similar environment where little or no cooling effect is revealed (as also reported in Fig. 4).

Finally, the burnishing speed plays a little role on the variation of burnishing forces although a slight increase with increasing the speed is observed. This tendency is due to the absence of the thermal softening phenomenon since very low temperature are reached during the process (Fig. 4). Therefore, the strain rate plays a significant role on the burnishing forces variation. Thus, the low temperature variation registered during all the investigated tests also implies that no dynamic recrystallization occurs during burnishing process at the selected process conditions.

3.2 *Mean Surface Roughness*

The mean surface roughness of the as machined sample was probed to be of about 1.32 μm. On the contrary, burnished surface roughness showed significant improvement reaching up to the 79% of reduction from the as machined surface. Figure 5 shows that with increasing of burnishing force the surface roughness decreases since surface irregularities can be reduced or eliminated by pressing the roller on the cylindrical surface.

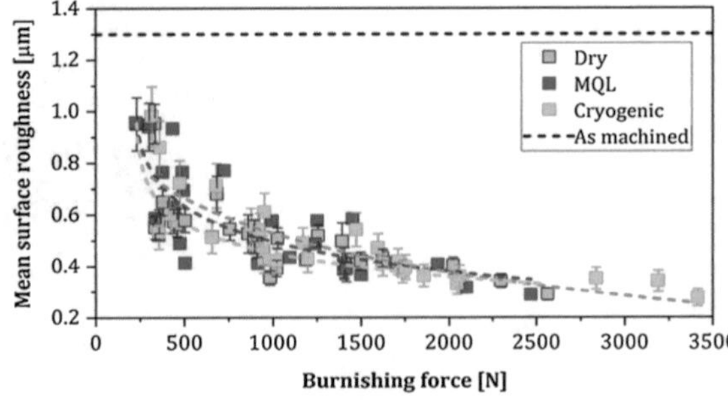

Fig. 5 Mean surface roughness (Ra) of burnished samples under dry, MQL and cryogenic cooling conditions (the error bars indicate data dispersion)

This implies a greater volume of compressed material during the process which, within the investigated range of parameters, leads to a greater portion of burnished surface undergoing significant flow filling more efficiently the asperities of the machined surface and smoothing the overall sample surface. Superior burnished surface roughness can also be achieved by using larger tool radius due to geometrical and operational effects.

In fact, a larger radius allows to press a higher amount of material under the tool as a consequence of a higher ploughing. Also, larger radius generates higher engage angle (Fig. 1d) which determines an increase of the burnishing force and, consequently, an improvement of the surface roughness. Finally, the cooling strategy shows a slight influence on the surface roughness since for a given burnishing force the Ra values are quite similar although MQL and dry burnishing seem to create superior surface quality. These evidences can be explained by the local rise of the temperature during dry burnishing. Thus, the surface asperities can be easily deformed while, the use of LN2 condition keeps the temperature range low reducing the ability of the material to flow during the process. Finally, since the MQL condition results to a similar environment of the dry (Fig. 4), thus the Ra values are quite similar to those obtained in dry condition.

3.3 *Microhardness and Microstructure*

A significant increase of surface hardness is revealed after burnishing process as shown in Fig. 5. The results data highlight that the hardness values increase with the burnishing forces as a consequence of a greater deformation layer depths and larger localized cold worked zones. Therefore, surface hardness values increase by selecting process parameters which lead to achieve higher burnishing force (Fig. 6). Also, the difference in cooling strategy was found to have a significant impact on the surface microhardness changes.

In particular, cryogenic cooling has been proved to greatly improve the strain hardening capability during high strain severe plastic deformation [13]. Thus, for a given burnishing force, cryogenic cooling produced superior microhardness compared to

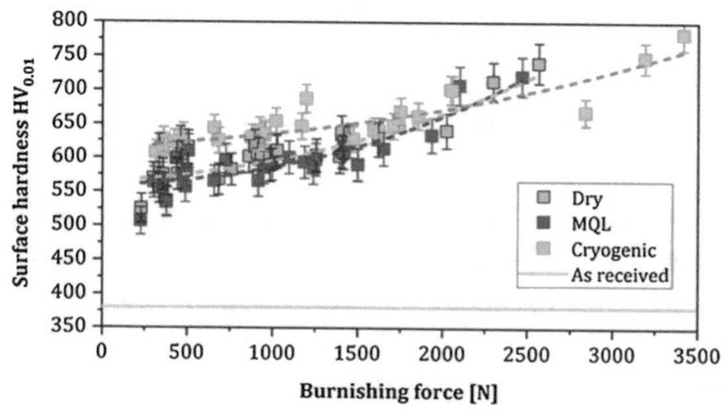

Fig. 6 Surface hardness at varying of burnishing force and cooling strategies (as received *Ti6Al4V* has a hardness value of 380 HV; the error bars indicate data dispersion)

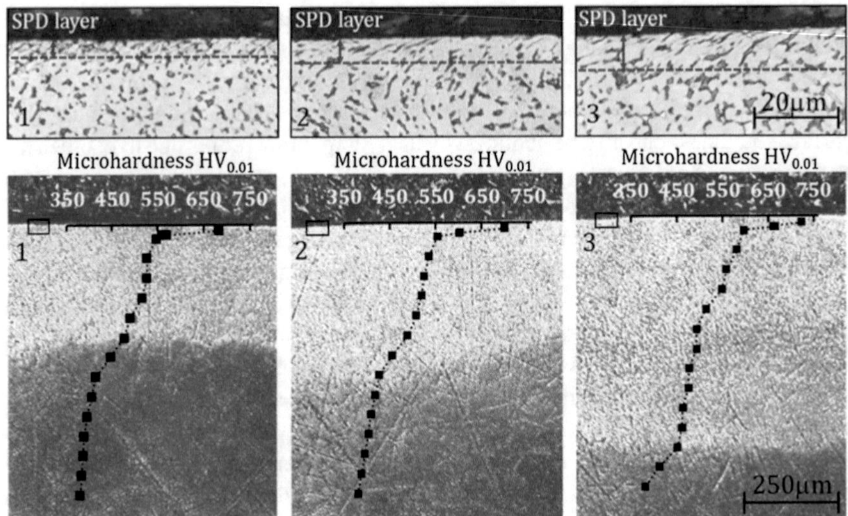

Fig. 7 Microstructure of samples burnished under **a** MQL, **b** dry and **c** cryogenic cooling conditions and sub-surface microhardness variation

the other lubrication strategies. In contrast, MQL and dry show comparable trends since the temperatures in both conditions are quite similar as previously explained. Therefore, similar burnishing force for a fixed process condition are produced. Figure 7 shows the influence on surface integrity when samples are burnished under MQL, dry, cryogenic conditions at speed of 200 m/min, burnishing depth of 0.5 mm and a roller radius of 5 mm.

In particular, the micrographs show the formation of a deep affected layer related to the hardness changes which gradually approaches to the bulk one, together with a severe plastic deformation layer (SPD) formed in the first few microns of the affected layer. The cryogenic is the cooling methodologies that shows the more significant hardness enhancement at deeper depth, while dry and MQL burnishing show an analogous affected layers. Therefore, for a given burnishing force, the SPD and affected layers in cryogenic burnishing are larger than those experimentally observed in MQL and dry burnishing (Fig. 8).

3.4 Wear Resistance

Generally, wear resistance can be improved in several manners [3] and within these surface modification processes, including burnishing, are able to induce positive effects on wear properties. In particular, (Fig. 9) an increasing of wear resistance (i.e., specific wear rate is reduced) can be observed by increasing the tool radius, the speed and the depth of burnishing.

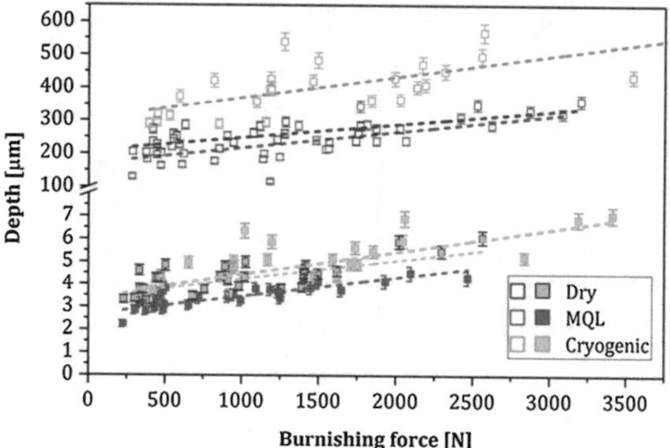

Fig. 8 SPD layer (filled symbols) and affected layer (empty symbols) at varying of burnishing force and cooling strategies (the error bars indicate data dispersion)

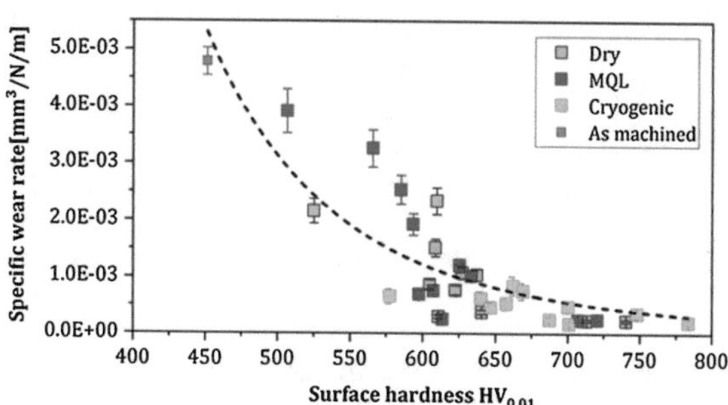

Fig. 9 Variation of wear rate with surface hardness

The above process parameters play a synergic effect in generating higher burnishing force (Fig. 3).

Consequently, the surface hardness increases with increasing the burnishing force (Fig. 6) resulting in a lower specific wear rate. The variation of the specific wear rate as function of the surface hardness is depicted in Fig. 9.

Also the affected layer significantly influences the wear resistance since the specific wear rate is lower for larger affected layer [3]. Therefore, the cryogenic burnishing produces the superior surface in terms of wear resistance since it is able to generate deeper affected layer and higher surface hardness values.

4 Overall Evaluation of Roller Burnishing of Ti6Al4V

The full factorial design of experiments allows to discuss the results highlighting a clear influence of the investigated process parameters and the variable of interests when burnishing Ti6Al4V. Figure 10 reports the main effects plot for the responses (i.e. burnishing force, mean surface roughness, affected layer, SPD layer, surface micro hardness and wear rate) versus the factors.

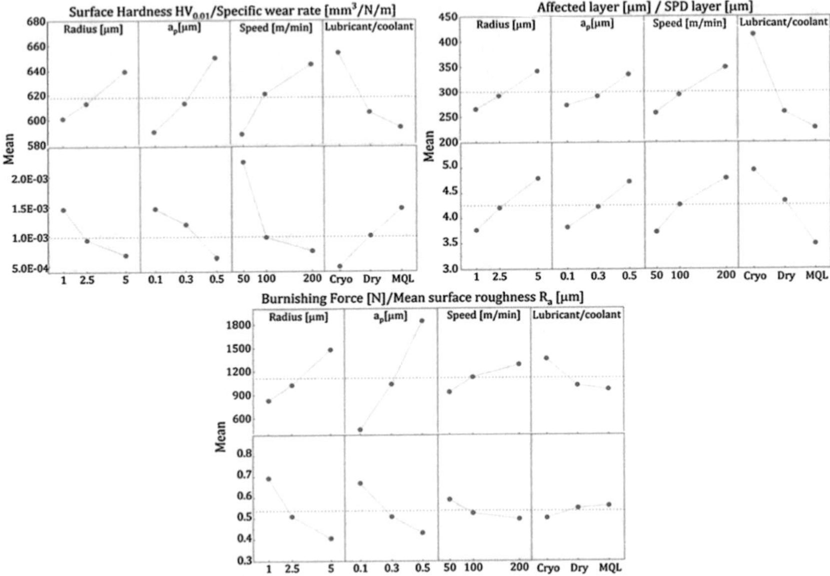

Fig. 10 Main effects plot for the complete full factorial experimental plan

The results summarize the overall variables trends for the investigated process parameters. In particular, the beneficial effect of the cryogenic cooling condition on the surface roughness, hardness and wear rate is clearly shown in Fig. 10.

Also, the affected layer is evidently deeper when the cryogenic cooling is applied and the initial SPD layer is more pronounced. These characteristics have been demonstrated to be beneficial since they also contribute to a deeper compressive residual stresses state on the surface [3, 4]. The main effect of the speed is to increase burnishing forces, decrease surface roughness and increase the surface hardness and affected layer depth. It also shows beneficial effect in reducing the specific wear rate. The same trend is also verified for the ranges of depth of burnishing and tool radii used during the experimental campaign.

5 Conclusions

The extensive experimental evidences reported in this paper show the influence of burnishing process parameters on surface quality and wear resistance of Ti6Al4V. The employed process parameters were able to generate a variety of burnishing force driving the surface modifications leading, in turn, to a change in tribological performance. The experimental evidences showed a trend where higher burnishing forces increased the surface hardness, reduced roughness and improved the wear resistance. The achieved results highlight the capability of the process to significantly improve the above performance especially when cryogenic cooling is applied. The MQL and dry also improve the wear performance of Ti6Al4V compared with the as machined one even if at a lower extent than the cryogenic one. From the other hand, when a drastic reduction of roughness is required, dry and MQL strategies offer the best results. Finally, the outcomes of this study allow to clearly find a combination of process parameters, within the investigated levels, able to offer the best surface quality according to the specific needs and applications of the Ti6Al4V components.

References

1. Jawahir IS, Brinksmeier E, M'Saoubi R, Aspinwall DK, Outeiro JC, Meyer D, Umbrello D, Jayal AD (2011) Surface integrity in material removal processes: recent advances. CIRP Ann Manuf Technol 60(2):603–626
2. Rotella G, Alfano M, Candamano S (2015) Surface modification of Ti6Al4V alloy by pulsed Yb-laser irradiation for enhanced adhesive bonding. CIRP Ann Manuf Technol 64(1):527–530
3. Schulze V, Bleicher F, Groche P, Guo YB, Pyun YS (2016) Surface modification by machine hammer peening and burnishing. CIRP Ann Manuf Technol 65(2):809–832
4. Caudill J, Schoop J, Jawahir IS (2018) Correlation of surface integrity with processing parameters and advanced interface cooling/lubrication in burnishing of Ti-6Al-4V alloy. Adv Mater Process Technol (in press)
5. Devaraya Revankar GD, Shetty R, Rao SS, Gaitonde VN (2017) Wear resistance enhancement of titanium alloy (Ti–6Al–4V) by ball burnishing process. J Mater Res Technol 6(1):13–32
6. Abrao AM, Denkena B, Breidenstein B, Mörke T (2014) Surface and subsurface alterations induced by deep rolling of hardened AISI 1060 steel. Prod Eng Res Dev 8:551–558
7. Brinksmeier E, Garbrecht M, Meyer D (2008) Cold surface hardening. CIRP Ann Manuf Technol 57(1):541–544
8. El-Taweel T, El-Axir M (2009) Analysis and optimization of the ball burnishing process through the Taguchi technique. Int J Adv Manuf Technol 41(3–4):301–310
9. Borgioli F, Galvanetto E, Iozzelli F, Pradelli G (2005) Improvement of wear resistance of Ti-6Al-4V alloy by means of thermal oxidation. Mater Lett 59:2159–2162
10. Luo H, Liu J, Wang L, Zhong Q (2006) Investigation of the burnishing force during the burnishing process with a cylindrical surfaced tool. J Eng Manuf Part B 220:893–904
11. Wang YM, Ma E, Valiev RZ, Zhu YT (2004) Tough nanostructured metals at cryogenic temperatures. Adv Mater 16(4):328–331
12. Wei Q, Cheng S, Ramesh KT, Ma E (2004) Effect of nanocrystalline and ultrafine grain sizes on the strain rate sensitivity and activation volume: fcc versus bcc metals. Mater Sci Eng 381(1):71–79
13. Jawahir IS, Attia H, Biermann D, Duflou J, Klocke F, Meyer D, Newman ST, Pusavec F, Putz M, Rech J, Schulze V, Umbrello D (2016) Cryogenic manufacturing processes. CIRP Ann Manuf Technol 65(2):713–736

Lead Time Analysis of Manufacturing Systems with Time-Driven Rework Operations

Alessio Angius and Marcello Colledani

Abstract Products whose quality features deteriorate with the time they spend in the manufacturing system are common in industry. Usually, they are perishable products that loose their performance or value whenever their sojourn time in the system exceeds a certain threshold. A typical example is the food industry where such products need to be identified and scrapped. In other cases, perishable products can be reworked for restoring their functions when their sojourn exceeds a threshold. For instance, automotive painting processes suffer of dust settling on parts before the drying process and semiconductor wafers risk contamination between the cleaning and heating processes. This work develops an analytical method to evaluate the performance of systems with rework operations triggered by an extended residence time. The approach is based on the exact computation of the lead time distribution in production lines featuring general markovian machines. By exploiting the developed method, specific production policies aiming at minimizing the rework rate and maximizing the effective throughput in such systems are investigated. The method is validated within a real case showing high benefits through industrial applications.

Keywords Lead-time · Rework · Time based policies

1 Introduction and Motivation

The production of deteriorating products must be regulated with strict delivery precision. This is because the quality of deteriorating products is strongly affected by their *lead time*, i.e. the time that they spend within the system or portion of it. Thus, time constraints must be set to avoid that products overstay in the system. Most of the

---

A. Angius (✉) · M. Colledani
Politecnico di Milano, Via La Masa 1, 20156 Milan, Italy
e-mail: angius@di.unito.it

M. Colledani
e-mail: marcello.colledani@polimi.it

E. Ceretti and T. Tolio (eds.), *Selected Topics in Manufacturing*, Lecture Notes in Mechanical Engineering,
https://doi.org/10.1007/978-3-030-57729-2_11

time, if the lead time of a product exceeds a certain threshold, the product has to be considered defected or, if not recoverable, perished (or obsolete).

A perished product must be scrapped by leading to financial losses. This is common in the food industry where, for example, delays before the pasteurization process might lead to the formation of mold. However, similar cases can be found in other contexts such as multi-lumen polymeric micro-tubes for medical vascular catheters where delays between the drying and the extrusion process might affect the quality of the tube section.

There are also situations where defective products might be recovered if they are re-inserted in the system and re-processed. This situation is common, for instance, in processes that are sensitive to dust deposits where cleaning must be performed again before moving the part to the next stage. Semi-conductor production is affected by this problem; delays in the sequence Implant-Cleaning-Oven generate rework if parts wait too long for the heating process in the ovens.

In order to deal with time constrained production, several researchers dedicated effort to provide methodologies that allow the analysis of these systems as well as the development of control strategies to improve system performances.

Some noticeable works in the context of perishable products are: Liberopoulos and Tsarouhas [1] where cost-efficient ways of speeding-up the croissant processing lines of Chipita International Inc. are reported, and Liberopoulos et al. [2] which focuses on the production rate of asynchronous production lines in which machines are subject to failures. In Wang et al. [3], the authors proposed a transient analysis to design the size of the buffers needed in dairy filling and packaging lines whereas in Subbaiah et al. [4] an inventory model for perishable products with random perishability and alternating production rate is proposed. The first work that deals explicitly with lead time is described in Shi and Gershwin [5] where the lead time between two machines having geometrically distributed failures and repairs is calculated. The work in Angius et al. [6] extends Shi and Gershwin [5] by considering general Markovian machines and providing a closed-form solution for the geometric case. The general Markovian machine model is used in Angius et al. [7] to analyze a real manufacturing system producing micro-catheters for medical applications. The same model is extended in order to analyze composed of N machines with different layouts (Lines, Closed Loop, Assembly) in Angius et al. [8] and for multi-product with dedicated buffers in Angius and Colledani [9]. Finally, lead time control policies to reduce the scrapping of parts are presented in Angius et al. [10].

For what concerns system with rework and, more specifically, semi-conductor production some notable works are the following: Kao et al. [11] which proposes an allocation and sequencing strategy based on a bottlenecks; Kao et al. [12] which describes a near optimal allocation strategy with batching and waiting time constraints for a furnace. Mathirajan and Sivakumar [13] that presents a review of scheduling of batch processors in semiconductor. Robinson and Giglio [14] that deals with the capacity planning for semiconductor wafer fabrication with time constraints; finally, in Scholl and Domaschke [15] a model to simulate the production of semi-conductor with time constraints is presented.

Despite the importance of the problem, at the best of our knowledge, literature does not offer stochastic models that describe explicitly the dynamics of systems having rework driven by lead time constraints.

This work represents a first attempt to fill this gap since it introduces an analytical method for the exact analysis of the lead-time distribution and system throughput for a system composed of N machines where parts deteriorate in a generic portion of the system, referred as *critical portion*. Deteriorated parts must be re-processed and are moved in a rework buffer that store the parts while they wait for being re-processed by the critical portion of the system.

The model is defined in such a way that machine dynamics are described by means of general Markov matrices. Therefore, the model is able to embed real behaviors. This makes the method suitable for being used in production integrated analysis as well as for providing insights that might suggest operational strategies.

The methodology is illustrated by means of experiments that show:

- how the lead time distribution and system throughput is affected by machine performances;
- the difference between the lead time distribution of a part that is processed for the first time and the distribution of reworked parts;
- how system performances are affected by the size of both the primary and the rework buffer.

Finally, a real case study based on semi-conductor production is presented.

The paper is structured as follows: Sect. 2 presents the system and the modeling assumptions; Sect. 3 illustrates the measures of interest; Sect. 4 illustrates in detail the numerical method used to compute the lead time distribution; Sect. 5 contains the numerical experiments and the analysis of a real case study; the conclusive remarks and future works are drawn in Sect. 6.

2 Modeling Assumptions

We consider discrete time manufacturing systems, i.e., we assume that time is divided into slots. The system is composed of N machines. As depicted in Fig. 1, machines are connected two-by-two by (N − 1) buffers that store the parts while they wait for

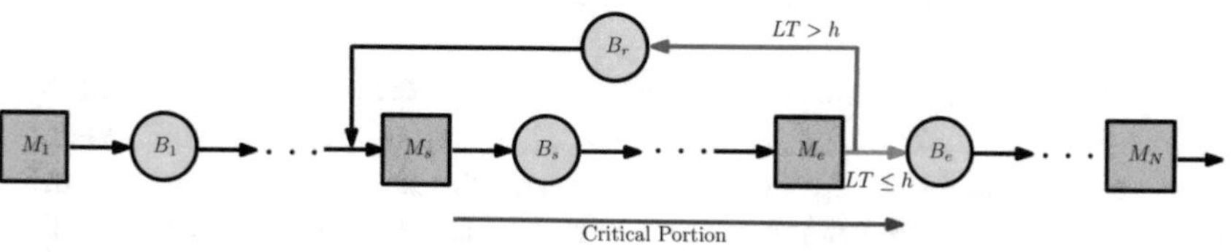

Fig. 1 Graphical representation of a system composed of N machines with time driven rework operations between machine M_s and machine M_e

being processed. As a consequence, machine M_i puts part on the ith buffer and $M_{(i+1)}$ removes them from it. An additional buffer, referred as *rework buffer*, is present to store the parts waiting for rework. The ith buffer has capacity B_i and the rework buffer has capacity B_r (in sake of simplicity, we will use the capacity of the buffers to identify them).

The system has a critical portion that starts with machine M_s, $1 \leq s < N$, and ends with machine M_e, $e > s$. The time that parts spend in this portion of the system, referred as *lead time*, cannot exceed a pre-fixed threshold, denoted with τ. When the lead time of a part is above τ, the part must be re-processed from all the machines that compose the critical portion. The tracking of the lead time always begins in the moment in which the parts ends its processing at machine M_s whereas there are two possible ending moments. The first is immediately before the service at M_e; in this case, the part that has exceeded the threshold is not processed by M_e but is instead moved to the rework buffer by allowing M_e to process another part (if available) on the buffer. On the contrary, if the service at M_e is included the part is always processed by the last machine of the critical portion and, if the lead time is above the threshold, the part is moved on the rework buffer. In this case, M_e is blocked if the rework buffer is full. We denote the first and the second situation, with e^- and e^+, respectively.

Each machine M_i, is characterized by a set of states S_i with dimensionality L_i.

The dynamics of each machine in these states are captured in a $L_i \times L_i$ probability matrix denoted with T_i. Moreover, a quantity reward vector μ_i is considered, with dimensionality L_i and binary entries: $\mu_{i,j} = 1$ if the machine is operational and it processes 1 part per time unit while in state j; if $\mu_{i,j} = 0$ the machine is down and it does not process parts in state j.

The states of machine M_i are partitioned according to μ_i into up states (the machine is operational), denoted as U, and down states (the machine is not operational), denoted as D. Without loosing generality, we assume that the states are ordered in such a way that T_i can be decomposed in the following four blocks:

$$T_i = \begin{bmatrix} \overline{P_i} & P_i \\ R_i & \overline{R_i} \end{bmatrix} \tag{1}$$

where, considering machine i, the block $\overline{P_i}$ contains the transition probabilities among the up states, $\overline{R_i}$ among the down states, P_i from up states to the down states (leading to a break down) and R_i from down states to up states (leading to repair).

The generic state indicator for this system is conveniently described by a tuple $s = (b_1, \ldots, \boldsymbol{b}_s, \ldots, \boldsymbol{b}_{e-1}, b_e, \ldots, b_{N-1}, b_r, \alpha_1, \alpha_2, \ldots, \alpha_N) \in S$, where α_i assumes values in the set S_i, the terms b_i describe the amount of parts in the buffers that do not belong the critical portion whereas the terms $\boldsymbol{b}_i$ refer to the buffers belonging to the critical portion. In particular, the latter are vectors of the form $\left(b_{i,1}, b_{i,2}, \ldots, b_{i,B_p}\right)$ where $b_{i,j}, k \in \left[1, B_j\right]$, exists in $\{E\} \cup [0, \tau]$ and describes the time spent in the critical portion by the part on the jth slot of the ith buffer up to the threshold. The term E is a placeholder required to describes empty slots.

In each time slot, the state of the machine is determined at the beginning of the time unit. The buffers are updated according machine services at the end of the time unit.

Machine M_1 is never starved and machine M_N is blocked only when $e^+ = N$ and B_r is full.

Machine M_s takes parts from B_r as long as they are present.

Operational Dependent Transitions are assumed, i.e. a machine cannot make transitions to other states if it is starved or blocked. Machine blocking is determined before the service (BBS mechanism).

3 Performance Measures

The problem tackled in this work can be formulated as follows. Given the system described by the assumptions detailed in the previous section, calculate the following performance measures:

- E: the system production rate of not defective parts.
- $P(LT_p = h)$: probability with which a part leaves the critical portion in h time units when is processed for the first time.
- $P(LT_r = h)$: probability with which a part leaves the critical portion in h time units when it has been reworked.
- $P(LT = h)$: probability with which a part leaves the critical portion in h time units independently from how many times it has been processed.
- P(R) $= P(LT > h)$: probability of rework.

4 Numerical Method

In this section we describe the method to calculate the distribution of the lead time from the moment in which a part ends enters the critical portion and the moment in which it goes out from it.

The method is composed of three steps. In the first step, the DTMC describing system dynamics is built and its steady-state is computed. The second step uses the steady state distribution of the DTMC to determine the initial condition for the computation of the lead time. Finally, the third step carries on the computation of the lead time by computing the time of absorption of the DTMC which has been modified in order to catch the moment in which parts leave the critical portion.

All the three steps are computed by using only matrix-vector operations.

4.1 Construction DTMC and Computation Steady-State

The construction of the DTMC describing the system can be done by applying the assumptions described in Sect. 2 and the rules described in Buchholz and Telek [16] where the rules necessary to enhance Markov processes with Phase-type distributed transitions are described.

The result is a DTMC described by a block-matrix where each not-null block is expressed in the form $\otimes_{i=1}^{N} A_i$ with $\otimes$ being the Kronecker product and A_i being the matrix that describes the change of state of machine i.

Each matrix A_i can assume five different values; the first four correspond to the internal transitions of the machine i as described in Eq. (1) whereas the fifth is an identity matrix (I) having dimension equal to $\overline{P_i}$. The identity matrix is used in place of matrix $\overline{P_i}$ whenever machine i is operative but blocked or starved. This is necessary to implement operational failures when a machine is not processing parts. In detail, we have that A_i assume values: (i) $\overline{P_i}$ and P_i if $a_i = U$ and B_{i-1} is not empty and B_i is not full; (ii) $\overline{R_i}$ and R_i if $a_i = D$; (iii) $A_i = I$ if $a_i = U$ but the upstream buffer is empty or the downstream buffer is full. Furthermore, $A_{e^+} = I$ also when the rework buffer is full.

Buffers are updated at the end of the time slots according to the states of the machines. Thus, we have that b_r is decremented by M_s whenever $\alpha_s = U$ and B_s is not full whereas it is incremented when buffer B_{e-1} contains a part that has exceeded the threshold (before or after the service at M_e). A slot of $\boldsymbol{b}_s$ will contain a part with sojourn time equal to 0 only if M_s will be up and operative at the end of the time slot and the oldest part of each buffer of the critical portion will be moved to the next buffer if the corresponding machine is up at the end of the time unit. Additionally, the sojourn times of the entries remaining on the buffers of the critical portion will be increased by one.

In order to better understand the dynamics, consider the two-machine case with $s = 1$ and $e^+ = 2$, $B_1 = B_r = 3$ and $\tau = 5$. Furthermore, consider state ((1,3,5),1,U,U). In this state, there are only two possible transitions because buffer B_1 is full and, consequently, M_1 is blocked. The first transition arrives at state ((E,2,4),2,U,U) although the two machines did not change state. In particular, the term (E,2,4) describes the fact that the oldest part has been removed from B_1 and the sojourn part of the two remaining parts is increased by one; the number of parts in B_r is increased by one because the part removed by M_2 had reached the threshold before being processed; Due to the fact that M_1 is blocked, the probability of this transition is given by the product $I \otimes \overline{P}_2$.

The other transition instead represents the case in which the downstream machine fails; thus, it occurs with probability $I \otimes P_2$. In this case, no part is moved and the only term that is updated is $\boldsymbol{b}_1$ because the sojourn time of the parts changes. As a consequence, the transition leads to state ((2,4,5),1,U,U). Note that the sojourn time of the oldest part in the buffer does not change because it has already reached the threshold. Four transitions are instead possible from state ((E,2,4),2,U,U) because M_1 can process parts; therefore, both the machines can fail or remain operative. For example, if both the machines remain operative, the next state will be ((E,0,3),1,U,U) because a new part enters B_1 and the one that is removed by M_2 leaves the system because it has not reached the threshold. The probability of this transition is $\overline{P}_1 \otimes \overline{P}_2$ because both the machines stay operative.

If not blocked, machine M_1 takes parts from outside the system when B_r is empty; thus, for example, it is possible to reach ((E,0,1),0,U,U) from state ((E,0,3),0,U,U) if the two machines remain operative. On the contrary, M_2 is blocked when a part in B_1 has reached the threshold and B_r is full, e.g. only the upstream machine can change its state in ((E,3,5),3,U,U) and $A_2 = I$ for both the transitions.

The dynamic of the DTMC are conveniently collected in a matrix Q whose steady state distribution, denoted by φ, can be computed by using both traditional and advanced numerical techniques such as those for quasi-birth-death processes.

4.2 Computation of the Initial Vector

The next step consists in computing the initial condition for the computation of the lead time i.e. the distribution of the lead time in h = 0. This corresponds in catching the moment in which a new part is placed on B_s. In order to compute the probability distribution of this event, we consider the steady state distribution and force the DTMC in making an additional jump by using only those transitions that insert a part on the B_s. These transitions can be selected directly from the matrix Q and the computation is carried on by using the following formula:

$$\varphi' = \varphi\left(F^{\langle cond_1 \rangle} Q F^{\langle cond_2 \rangle}\right)$$

where $F^{\langle cond_1 \rangle}$ and $F^{\langle cond_2 \rangle}$ are filtering matrices whose entries are equal to one on the diagonal only if the logical expressions $\langle cond_1 \rangle$ and $\langle cond_2 \rangle$ are true in the corresponding state; otherwise, they are equal to zero. The logical expression $\langle cond_1 \rangle$ determines the states where part can be processed by M_s whereas $\langle cond_2 \rangle$ discriminate from those states in which the machine fails during the processing.

The focus of the analysis determines the first condition. If the analysis does not discriminate between parts that enter the critical portion for the first time and rework parts, then $\langle cond_1 \rangle$ checks only that B_s is not full. On the contrary, if the analysis focuses on parts that enter the system for the first time, $\langle cond_1 \rangle$ must verify also that the rework buffer is empty. Vice versa if the analysis is focusing on reworked parts, $\langle cond_1 \rangle$ selects those transitions in which M_s machine is not blocked and the rework buffer is not empty.

Since the blocking of machine M_s is verified at the beginning of the time slot, $\langle cond_2 \rangle$ verifies only that the machine is up at the end of the time unit. Thus, $\langle cond_2 \rangle = M_s == \text{U}$.

The vector $\varphi^{'}$ does not sum to one and requires a normalization to be a proper distribution. This is done by dividing the vector by is norm-1. As a consequence, we have that the initial condition of the lead time as $\pi(0) = \varphi^{'}/|\varphi^{'}|$.

Note that, by definition, $|\varphi^{'}|$ corresponds to the system throughput if we are not discriminating parts or to the portion of throughput that is generated by new parts and rework if the analysis is focusing on only one of the two.

4.3 Computation of the Lead Time

The lead time is a duration that starts from the moment in which an item is put on the B_s by the upstream machine and ends in the moment in which the part is removed from B_{e-1} (before or after the service). In order to compute the lead time, we must track the last part inserted on the buffer for all the possible buffer levels. By defining $\pi(0)$, we caught the moment in which the part entered. In order to catch the removal, we need to catch the moment in which the part leaves the portion.

This is done by building a modified DTMC in which M_s still removes parts from B_r but it is inhibited from putting them on B_s. This guarantees that all the buffers of the critical portion will eventually get empty. Furthermore, since the system uses a FIFO policy, the last part that will leave the buffers of the critical portion will be the one whose insertion was forced in the previous step.

The states of the system where all the buffers of the critical portion are empty are the absorbing states. The moment in which the DTMC reaches these states corresponds to the removal of the tracked part.

Let us denote the absorbing DTMC with $\overline{Q}$, the computation of lead time over time follows the relation: $\pi(h) = \pi(h-1)\overline{Q}$.

The cumulative distribution of the lead time is derived by summing the probability to find the process in one of the absorbing states. Let $empty(s)$ be a function that determines if the buffers of the critical portion are empty in state s and let $\pi_{s,k}(h)$ be the probability of the kth entry of the block corresponding to state s of the absorbing DTMC. Then, we have that:

$$P(LT \leq h) = \sum_{\forall s: empty(s)=True} \sum_{\forall k \in s} \pi_{s,k}(h) \tag{2}$$

The construction of $\hat{}\overline{Q}$ follows the same rules that have been used to build Q.

5 Experiments

This section reports the numerical results obtained by applying the theoretical framework presented in this paper.

In the next paragraphs, we consider systems having two machines connected by one buffer and the rework buffer.

The decision of limiting the experiments to two machines has been taken in order to facilitate the understanding of the results. The analysis of longer lines introduces additional complexity that is difficult to describe properly in a limited number of pages. Authors believe that being able to show interesting and counterintuitive results in a simplified scenario put on the spotlight the relevance of the problem and paves the way to further analysis that will be driven by the results presented in the following.

The first set of experiments have been performed by using machines having geometrical failures and repairs.

Subsequently, we present a real case from semi-conductor industry in order to illustrate the practical implications of our work and the potential improvements that can be achieved by using the method to support everyday operations and design choices.

The proposed method is exact. Therefore, results do not need to validation against Monte Carlo simulation. Each single computation of the numerical method presented in the previous section took less than a second by using a JAVA prototype on a Centrino I7 with 16 Gb of RAM.

5.1 Balanced System with Geometric Up-Down Machines

This section focuses on a system composed of two single-failure machines having the same efficiency. In particular, the two machines have geometric failures and repairs with average 100 and 10 time units, respectively. Both buffers B_1 and B_r have capacity equal to 10.

The first case considers $s = 1$ and $e^+ = 2$; thus, the lead time includes the processing of the parts at the second machine. The experiment aims to investigate the impact of different values of τ on the distribution of the lead time. In particular, we tested:

- $\tau = 5$, which is a tight time constraint in comparison with the buffer capacity B_1; in fact, parts that enter the buffer after the 5th slot are guaranteed to be re-processed;
- $\tau = 10$, which corresponds to an optimistic threshold because the probability with which parts can cross the system only once is not null for all the slots in the buffer but only if the downstream machine does not fail while during their sojourn;
- $\tau = 15$, which is a time interval able to compensate short downtimes of the downstream machine.

We analysed the lead time for all the three cases. The analysis has been performed with and without making the distinction between parts that are entering the system for the first time and reworked parts. Results are reported in Fig. 2.

All the three cases show that the lead time distribution is characterized by a bi-modal behaviour where peaks are placed at 1 and $B_1 - 1$. The peaks reflect the two most dominant slots on the buffer: the first and last. By making a comparison between

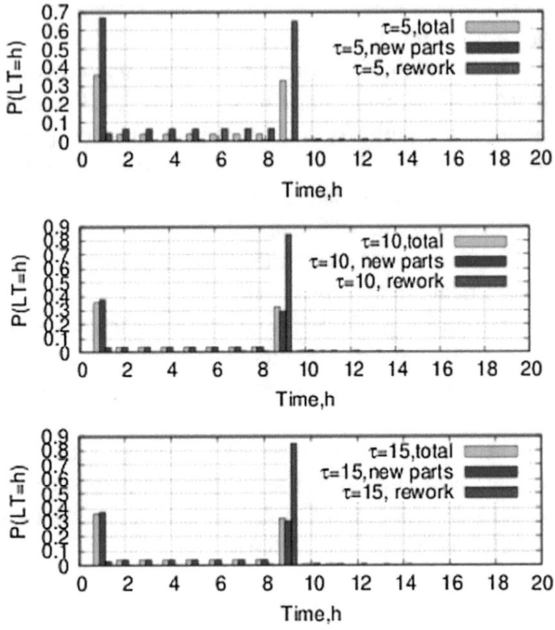

Fig. 2 Comparison of lead time distributions with $s = 1$ and $e^{+} = 2$ for the case with two-machines having geometrical failures and repair with parameters $p = 0.01$ and $r = 0.1$, buffer capacities $B_1 = B_r = 10$, and different thresholds τ

the lead time distribution of a part that enters the system for the first time and with the distribution of a reworked part we can observe that the two distributions have a symmetrical behaviour. The parts processed for the first time are more likely to go out from the critical section in a short time interval since the major part of the distribution is located around small values. On the contrary, reworked parts are more likely to stay longer in the system because the probability mass of their lead distribution is placed mostly around large values. This phenomenon can be explained by observing that parts exceed the threshold when they are placed in one of the last slot on the buffer. This situation is generated by a failure of M_2. When finally M_2 returns operative, it starts removing parts from B_1 and puts them on B_r. Consequently, as soon as B_1 has a slot available, Machine M_1 starts removing parts from the rework buffer to place them again on B_1 where they will occupy once again one of the last slots. Thus, as long as M_1 does not fail, buffer B_1 will not get empty and reworked parts will be always placed on the last slots available by increasing their probability to be re-processed.

By comparing the plots in Fig. 2, it is possible to notice that the more τ is large and the more the lead time distribution of new parts becomes alike to the overall lead time distribution while the distribution of reworked parts diverges and is almost completely placed at time $B_1 - 1$. This is a consequence of the fact that the larger is the threshold and the rarer is the rework. Therefore, the case in which a new part enters the system and goes out without being reworked is the dominant situation. On the other hand, rework becomes an exceptional situation that is not representative of the average system behaviour.

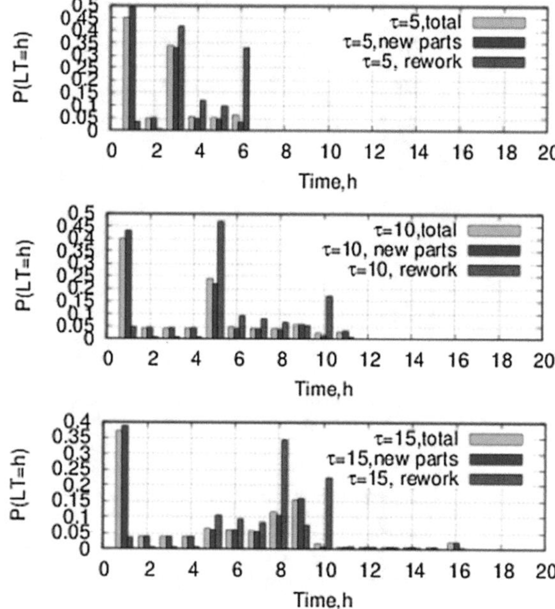

Fig. 3 Comparison of lead time distributions with $s = 1$ and $e^- = 2$ for the case with two-machines having geometrical failures and repair with parameters $p = 0.01$ and $r = 0.1$, buffer capacities $B_1 = B_r = 10$ and different thresholds τ

The second case takes in consideration the same system but the critical portion ends before the service at M_2. Formally, $s = 1$ and $e^- = 2$.

In this situation, deteriorated parts are removed from B_1 as soon as they are the oldest part on B_1 without being processed by M_2. Figure 3 provides all the possible lead distributions by using the same thresholds of the first case.

We can observe that the mass of the distributions tends to be located around smaller values in comparison with the previous experiment. Furthermore, the distributions are characterized by an additional peak that shifts on the r.h.s. for larger values of τ. This is consistent with the fact that, by increasing the threshold, parts are allowed to stay longer on the buffer. As last, the part of the distribution between the peaks is irregular and with marked differences between different time points; this it is far from the equiprobabilty that characterized the previous case. This phenomenon points out that, in this case, also the slots in the middle of the buffer strongly affect the lead time distribution.

The third case focuses on the system throughput; specifically, it investigates how the capacities B_1 and B_r affect the rework and, consequently, penalize the system performances. For this case, we assumed $\tau = 7$ and tested both the critical portions. Figure 4 depicts the results by putting on the spotlight how the increasing of B_1 is eventually detrimental for the performance of the system.

This is evident for the case with critical portion with boundaries $s = 1$ and $e^+ = 2$. In this case, the throughput drops for $B_1 \geq \tau$. The system with critical portion with boundaries $s = 1$ and $e^- = 2$ is instead more resistant to the decay of the throughput at the increasing of B_1. In particular, the only curve that drops is the case $B_r = 1$

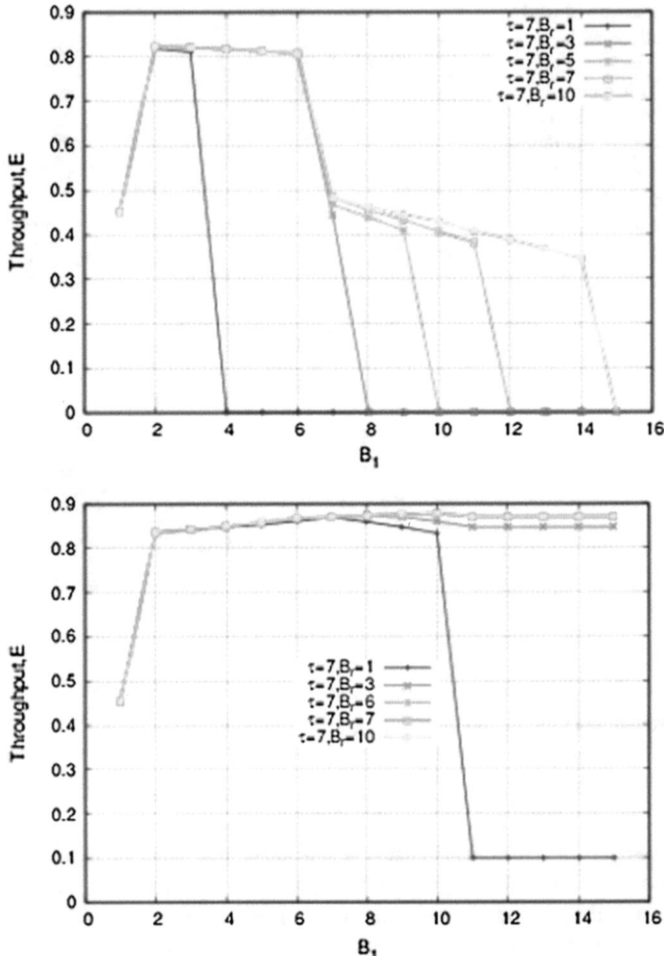

Fig. 4 System throughput of a two-machine system having geometrical failures and repair with parameters $p = 0.01$ and $r = 0.1$ as function of the buffer capacities B_1 and B_r by considering a threshold $\tau = 7$ and two different critical portions: $s = 1$ and $e^+ = 2$ (left)

whereas the others have a single maximum point and then they decrease to flatten around a single value (referred as *plateau* from now-on).

This resistance is consequence of the removal from the buffer that guarantees that M_2 does not waste cycle times to process parts that will be reworked anyway. The drop for the case $B_r = 1$ is instead consequence of the heavy blocking caused by the rework buffer.

The obtained results suggest that whenever possible, the processing of deteriorated parts should be avoided.

At the same time, results suggest that the increasing of the capacity of the rework buffer is always beneficial for the system performances because system blocking reduces if B_r is increased. On the contrary, large value of B_1 might be detrimental for the system because, even if the system throughput stabilizes around a value, the optimal throughput might be different.

5.2 *Case Study*

The last case is based on a real industrial scenario. In particular, we take in consideration the production of semi-conductors at the Bosch factory in Reutlingen. More specifically, we consider the *diffusion area* which is usually the most critical area in semi-conductor factories. In this area, a number of wafers is placed in a cylindrical reactor, which is then sealed, heated, and filled with a carrier gas to allow dopant atoms present in the gas to diffuse into the exposed layer of the wafers, altering their electrical and chemical characteristics [11]. Wafers are processed in standard lots whose size is dictated by material handling considerations. It is possible to process a number of lots together as a batch. Once the processing of a batch has been initiated, no jobs can be removed or added to the batch until it completes its processing. Due to the chemical nature of the process, it is impossible to process jobs with different recipes together in the same batch, and all jobs with the same recipe require the same processing time.

After the cleaning process there is a defined time frame in which the batches must enter the ovens. If this time frame is exceeded the wafers have to go again through the cleaning process.

For the purposes of this experiment, we took in consideration fixed batch sizes and we analyzed the processing times and the reliability of the machines. We focused on the Cleaning and heating stages performed by the Ovens.

Data has been extracted directly from the data log of the system. By using the fitting method described in Horváth et al. [17], we were able to build two machine models able to satisfy the requirement of Eq. (1) and maintain the statistical properties of failures, repairs and processing times of the machines.

The lead time of the parts is controlled before entering the oven; if their lead time is above time units, they are moved to the rework buffer. Therefore, the deteriorated batches never enter the oven. As a consequence, the critical portion of the system is $s = 1$ and $e^- = 2$.

We considered three different types of batches. The first deteriorates after they have spent 8 time units in the buffer whereas the second and the third deteriorates after 12 and 4 time units, respectively.

We evaluated the same model by searching of the optimal value B_1^{opt} i.e. the value of B_1 which provides the highest system throughput. The rework buffer has been instead considered fixed because evidences from the plant show that it does not play a crucial role in this scenario. Therefore, we set B_r in such a way that the blocking of parts waiting for rework was negligible.

Table 1 Percentage of error between optimal, minimal and plateau throughput of the diffusion area between the cleaning and heating stages by considering three different type of wafers

Batch type	τ (time units)	Δ% optimal/min	Δ% optimal/plateau	Δ% plateau/min
1	8	57.14101	4.636767	55.05711
2	12	43.38253	10.17258	36.97084
3	4	76.57486	8.57477	74.37782

For all the three cases, the analysis has shown a behavior similar to the one depicted on the right plot of Fig. 3, i.e. the system throughput as function of B_1 has a single the maximum point and then flattens on a plateau that is maintained all along. The minimum system throughput was instead always found for $B_1 = 1$. The percentage of error between optimal, minimal and plateau throughput are reported in Table 1.

Results show that the theoretical improvement is relevant. In particular, we have that B_1^{opt} provides an improvement of at least 40% in comparison with the minimum and at least 4% in comparison with the plateau. The most significant gap between optimal solution and plateau is found for the second type where there is an improvement greater than 10%. The largest gap with the minimum is instead found for the third type, ~76%.

6 Conclusive Remarks and Future Works

This article proposes for the first time the analysis of systems with time-driven rework operations. The method can be used to analyse production lines of arbitrary length. Furthermore, the method allows the use of machine models embedding dynamics that are robust from a statistical point of view. Therefore, the method is suitable for being used on real problems and everyday operative scenarios. Numerical results prove that system parameters, such as buffer capacities, strongly affect the system performances when the lead time of the parts is bounded by a threshold by leading to counterintuitive behaviours such as the drop of the throughput in response to the increment of the buffer capacity. Numerical results showed also that, whenever possible, the removal of deteriorated parts from the buffers should be anticipated in order to not waste cycle times.

This paves the way to investigate the definition of control policies and removal strategies for the deteriorated parts on the buffers.

The relevance of the method has been proved also by means of on a real case study in the context of semi-conductor production. The case study has shown that significant performance improvements might be achieved. In particular, a theoretical increment of at least 4% of the system throughput can be achieved by reducing the buffer capacity of the critical portion.

Future research will be targeted toward the extension of the proposed method in such a way that different system layout can be considered. Additionally, effort will be dedicated to improve the numerical method in order to speed up the computation for the cases where the critical portion is large. Finally, more complex dynamics as well as different deterioration types will be investigated.

References

1. Liberopoulos G, Tsarouhas PH (2002) Systems analysis speeds up Chipita's food processing line interface. Interfaces
2. Liberopoulos G, Kozanidis G, Tsaroulas PH (2007) Performance evaluation of an automatic transfer line with WIP scrapping during long failures. Manuf Serv Oper
3. Wang J, Hu Y, Li J (2010) Transient analysis to design buffer capacity in dairy filling and packing production lines. J Food Eng
4. Subbaiah KV, Rao S, Rao K (2011) An inventory model for perishable items with alternating rate of production. Int J Adv Oper Manag
5. Shi C, Gershwin S (2012) Part waiting time distribution in a two-machine line. In: Symposium on information control problems
6. Angius A, Colledani M, Horvath A (2014) Lead time distribution in unreliable production lines processing perishable products. In: ETFA 2014. IEEE, Barcelona
7. Angius A, Colledani M, Horvath A (2015a) Production quality performance in manufacturing systems processing deteriorating products. CIRP Ann Manuf Technol
8. Angius A, Colledani M, Horvath A, Gershwin S (2015b) Lead time dependent product deterioration in manufacturing systems with serial, assembly, and closed-loop layout. In: SMMSO, Volos
9. Angius A, Colledani M (2018) Analysis of the lead time distribution in multi-product systems with dedicated buffers. In: MIM'18, Bergamo
10. Angius A, Colledani M, Horvath A (2017) Lead-time oriented production control policies in two-machine production lines. IISE Trans
11. Kao YT, Zhan SC, Chang SC (2013) Bottleneck-centric pull and push allocation and sequencing of wet-bench and furnace tools. In: e-Manufacturing & design collaboration symposium (eMDC). IEEE
12. Kao YT, Zhan SC, Chang SC, Ho JH, Wang P, Luh PB (2011) Near optimal furnace tool allocation with batching and waiting time constraints. In: Automation science and engineering (CASE). IEEE
13. Mathirajan M, Sivakumar A (2006) A literature review, classification and simple meta-analysis on scheduling of batch processors in semiconductor. Int J Adv Manuf Technol
14. Robinson JK, Giglio R (1999) Capacity planning for semiconductor wafer fabrication with time constraints between operations. In: 31st conference on winter simulation: simulation—a bridge to the future. ACM
15. Scholl W, Domaschke J (2000) Implementation of modeling and simulation in semiconductor wafer fabrication with time constraints between wet etch and furnace operations. IEEE
16. Buchholz P, Telek M (2010) Stochastic Petri nets with matrix exponentially distributed firing times. Perform Eval
17. Horváth G, Buchholz P, Telek M (2005) A MAP fitting approach with independent approximation of the inter-arrival time distribution and the lag correlation. In: Conference on the quantitative evaluation of systems (QEST'05)